Thermospheric Circulation

**Progress in
Astronautics and Aeronautics**

Martin Summerfield,
Series Editor
PRINCETON UNIVERSITY

VOLUMES

EDITORS

1. **Solid Propellant Rocket
   Research. 1960**

Martin Summerfield
PRINCETON UNIVERSITY

2. **Liquid Rockets and
   Propellants. 1960**

Loren E. Bollinger
THE OHIO STATE UNIVERSITY
Martin Goldsmith
THE RAND CORPORATION
Alexis W. Lemmon Jr.
BATTELLE MEMORIAL INSTITUTE

3. **Energy Conversion for
   Space Power. 1961**

Nathan W. Snyder
INSTITUTE FOR DEFENSE ANALYSES

4. **Space Power Systems. 1961**

Nathan W. Snyder
INSTITUTE FOR DEFENSE ANALYSES

5. **Electrostatic Propulsion. 1961**

David B. Langmuir
SPACE TECHNOLOGY LABORATORIES, INC.
Ernst Stuhlinger
NASA GEORGE C. MARSHALL SPACE
FLIGHT CENTER
J. M. Sellen Jr.
SPACE TECHNOLOGY LABORATORIES

6. **Detonation and Two-Phase
   Flow. 1962**

S. S. Penner
CALIFORNIA INSTITUTE OF TECHNOLOGY
F. A. Williams
HARVARD UNIVERSITY

7. **Hypersonic Flow Research.
   1962**

Frederick R. Riddell
AVCO CORPORATION

8. **Guidance and Control. 1962**

Robert E. Roberson
CONSULTANT
James S. Farrior
LOCKHEED MISSILES AND SPACE
COMPANY

9. **Electric Propulsion
   Development. 1963**

Ernst Stuhlinger
NASA GEORGE C. MARSHALL SPACE
FLIGHT CENTER

10. **Technology of Lunar
    Exploration. 1963**

Clifford I. Cummings and
Harold R. Lawrence
JET PROPULSION LABORATORY

11. **Power Systems for Space Flight. 1963**

Morris A. Zipkin and
Russell N. Edwards
GENERAL ELECTRIC COMPANY

12. **Ionization in High-Temperature Gases. 1963**

Kurt E. Shuler, Editor
NATIONAL BUREAU OF STANDARDS
John B. Fenn, Associate Editor
PRINCETON UNIVERSITY

13. **Guidance and Control — II. 1964**

Robert C. Langford
GENERAL PRECISION INC.
Charles J. Mundo
INSTITUTE OF NAVAL STUDIES

14. **Celestial Mechanics and Astrodynamics. 1964**

Victor G. Szebehely
YALE UNIVERSITY OBSERVATORY

15. **Heterogeneous Combustion. 1964**

Hans G. Wolfhard
INSTITUTE FOR DEFENSE ANALYSES
Irvin Glassman
PRINCETON UNIVERSITY
Leo Green Jr.
AIR FORCE SYSTEMS COMMAND

16. **Space Power Systems Engineering. 1966**

George C. Szego
INSTITUTE FOR DEFENSE ANALYSES
J. Edward Taylor
TRW INC.

17. **Methods in Astrodynamics and Celestial Mechanics. 1966**

Raynor L. Duncombe
U.S. NAVAL OBSERVATORY
Victor G. Szebehely
YALE UNIVERSITY OBSERVATORY

18. **Thermophysics and Temperature Control of Spacecraft and Entry Vehicles. 1966**

Gerhard B. Heller
NASA GEORGE C. MARSHALL SPACE FLIGHT CENTER

19. **Communication Satellite Systems Technology. 1966**

Richard B. Marsten
RADIO CORPORATION OF AMERICA

20. **Thermophysics of Spacecraft and Planetary Bodies**
Radiation Properties of Solids and the Electromagnetic Radiation Environment in Space. 1967

Gerhard B. Heller
NASA GEORGE C. MARSHALL SPACE FLIGHT CENTER

21. **Thermal Design Principles of Spacecraft and Entry Bodies. 1969.**

Jerry T. Bevans
TRW SYSTEMS

22. **Stratospheric Circulation. 1969**

Willis L. Webb
ATMOSPHERIC SCIENCES LABORATORY, WHITE SANDS, AND UNIVERSITY OF TEXAS AT EL PASO

23. **Thermophysics: Applications to Thermal Design of Spacecraft. 1970**

Jerry T. Bevans
TRW SYSTEMS

24. **Heat Transfer and Spacecraft Thermal Control. 1970**

John W. Lucas
JET PROPULSION LABORATORY

25. **Communication Satellites for the 70's: Technology. 1971**

Nathaniel E. Feldman
THE RAND CORPORATION
Charles M. Kelly
THE AEROSPACE CORPORATION

26. **Communication Satellites for the 70's: Systems. 1971**

Nathaniel E. Feldman
THE RAND CORPORATION
Charles M. Kelly
THE AEROSPACE CORPORATION

27. **Thermospheric Circulation**

Willis L. Webb
ATMOSPHERIC SCIENCES LABORATORY, WHITE SANDS, AND UNIVERSITY OF TEXAS AT EL PASO

(Other volumes are planned.)

The MIT Press
Cambridge, Massachusetts,
and London, England

Progress in
Astronautics and Aeronautics

An American Institute of Aeronautics
and Astronautics Series

Martin Summerfield, Series Editor

Volume 27

Thermospheric Circulation

ı\

Edited by

Willis L. Webb

ATMOSPHERIC SCIENCES LABORATORY,
U. S. ARMY ELECTRONICS COMMAND,
WHITE SANDS MISSILE RANGE,
NEW MEXICO
AND PHYSICS DEPARTMENT, UNIVERSITY
OF TEXAS AT EL PASO

Library of Congress Cataloging in Publication Data
Main entry under title:

Thermospheric circulation.

(Progress in astronautics and aeronautics, v. 27)
"Selected papers presented at the summer institute
titled Physics of the Upper Atmosphere, which was conducted by the Physics Department of the University
of Texas at El Paso in cooperation with the Atmospheric
Sciences Laboratory, U.S. Army Electronics Command,
White Sands Missile Range during August 3-14, 1970."
Includes bibliographical references.
1. Atmospheric circulation--Addresses, essays,
lectures. 2. Atmosphere, Upper--Addresses, essays,
lectures. 3. Meteor trails--Addresses, essays,
lectures. 4. Radio meteorology--Addresses, essays,
lectures. I. Webb, Willis L., ed. II. Texas.
University at El Paso. Physics Dept. III. Atmospheric Sciences Laboratory. IV. Series.
TL507.P75 vol. 27 [QC880]   551.5'17   79-37738
ISBN 0-262-23053-4

Preface   xv

Foreword   xviii

1.   Upper Atmospheric Dynamics   1
        Willis L. Webb

1.1   Introduction   1
1.2   Electrodynamic Processes   5
1.3   Neutral Upper Atmospheric Dynamics   12
1.4   Neutral-Electrical Interactions   17
1.5   Dynaspheric Observational Program   25
1.6   References   29

2.   Lower Ionospheric Structure   37
        James E. Midgley

2.1   Electron Production   37
2.2   Electron Loss   42
2.3   Transport of Ionization   48
2.4   Time Variations   51
2.5   References   52

3.   The Topside Ionosphere   53
        Walter J. Heikkila

3.1   Introduction   53
3.2   The Geomagnetic Field   53
3.3   Motion of Charged Particles in Electromagnetic Fields   57
3.4   Electromagnetic Waves in a Plasma   60
3.5   Latitudinal and Longitudinal Ionospheric Variations   63
3.6   The Magnetosphere   67
3.7   Trapped Particles   68
3.8   The Outer Magnetospheric Region   70
3.9   Ionospheric and Auroral Phenomena in Relation to the
      Magnetosphere   71
3.10 Soft Particle Measurements   74
3.11 References   76

4.   Atmospheric Gravity Waves in Outline   79
        Colin O. Hines

4.1   General Nature   79

4.2 Governing Equations for the Elementary Case 79
4.3 Elementary Plane-Wave Solutions 81
4.4 Complications of Nonlinearity 84
4.5 Complications of Instabilities 85
4.6 Complications of Molecular Dissipation 85
4.7 Complications of Temperature Structure 86
4.8 Complications of Wind Structure 88
4.9 Complications of Earth Curvature and Rotation; Tides 89
4.10 The Semidiurnal Tide 90
4.11 The Diurnal Tide 90
4.12 Bibliography 92

5. Noctilucent Clouds — Their Characteristics and Interpretation 95
Benson Fogle

5.1 Introduction 95
5.2 The Height 98
5.3 Latitudes of Observation 98
5.4 Seasonal Distribution 99
5.5 Spatial Extent 99
5.6 Duration 100
5.7 Drift Motion 100
5.8 Wave Structure 101
5.9 Thickness and Vertical Wave Amplitude 101
5.10 Diurnal Variation 101
5.11 Auroral Influence 102
5.12 Year to Year Variation 102
5.13 Particle Size and Number Density 104
5.14 Theory 104
5.15 References 105

6. Noctilucent Cloud Wave Structure 109
B. Haurwitz

6.1 Introduction 109
6.2 The Wave Observations 109
6.3 Theoretical Speculations 111

| 6.4 | Suggestions for Further Studies | 114 |
| 6.5 | References | 115 |

7. **Meteor Trail Radar Winds over Europe** 117
Andre Spizzichino

7.1	Introduction	117
7.2	Description of the Equipment	118
7.3	Calibration and Errors	122
7.4	Hourly Rate of the Obtained Data-Comparison with Other Experiments	132
7.5	Data Processing	135
7.6	Experimental Results	156
7.7	Conclusions	176
7.8	References	177

8. **Radio Meteor Winds in the Southern Hemisphere** 181
R. G. Roper

8.1	Introduction	181
8.2	Observational Technique	182
8.3	Data Analysis	188
8.4	Prevailing Winds	192
8.5	Diurnal Variations	193
8.6	Semidiurnal Variations	195
8.7	Terdiurnal Variations	195
8.8	Wind Variability	196
8.9	Turbulence	197
8.10	References	201

9. **Radar Observations of Meteor Winds above Illinois** 205
M. D. Grossi, R. B. Southworth, and S. K. Rosenthal

9.1	Introduction	205
9.2	Considerations on Monostatic Versus Multistatic Radar Systems	208
9.3	Scattering Properties of a Meteor Trail	209
9.4	Position of the Meteor Trail	210
9.5	Wind Measurements	211
9.6	The SAO Phase-Coherent Radar System	213
9.7	Outlying/Remote Station	215

9.8	The Meteor Trail Radar	217
9.9	Format of the Raw Data and Example of Digital Recording	231
9.10	Foundations of the Method	234
9.11	Wind Profiles from Individual Meteors	240
9.12	Wind Field Computed from Many Meteors	244
9.13	Three-Dimensional Wind Measures	245
9.14	Expected Wind-Measurement Accuracies	246
9.15	Examples of Results	247
9.16	References	248
10.	**The Stanford Meteor Radar System**	**249**
	A. M. Peterson and R. Nowak	
10.1	Introduction	249
10.2	The Stanford Meteor Radar System Design Criteria	249
10.3	Application of the Stanford Radar System	256
10.4	Conclusions	259
10.5	References	260
11.	**Interactions between the Neutral Atmosphere and the Lower Ionosphere**	**261**
	C. F. Sechrist, Jr.	
11.1	Introduction	261
11.2	Evidence for Meteorological Influences on the Lower Ionosphere	262
11.3	Aeronomy of the Upper D Region	275
11.4	Theories of Neutral Atmosphere-Lower Ionosphere Interactions	310
11.5	Recommendations for Future Studies	315
11.6	References	317
12.	**Photochemical Models in Upper Atmospheric Research**	**327**
	Eigil Hesstvedt	
12.1	Introduction	327
12.2	Computational Methods	327
12.3	The Photochemical Model	328

13.   Airglow   331
       Rufus E. Bruce

13.1   Introduction   331
13.2   Prominent Spectral Features   331
13.3   Airglow Emission Brightness   334
13.4   OH Contamination   334
13.5   Airglow *Excitation Processes*   334
13.6   Airglow Morphology   341
13.7   Airglow as an Ionospheric Probe   346
13.8   References   347

14.   White Sands Missile Range Meteor Trail Radar Design   355
       Alton A. Duff

14.1   Introduction   355
14.2   The MTR System   355
14.3   Data Handling System   356
14.4   Calibration Equipment   358
14.5   Advanced Meteor Trail Radar   360
14.6   Advanced Digital Controller   361
14.7   Nike-Met Radar   361
14.8   Facility Layout   363
14.9   Conclusion   365
14.10 References   367

15.   Meteor Trail Radar Data Processing   369
       Bruce T. Miers

15.1   Introduction   369
15.2   Immediate Data Exchange Format   369
15.3   Data Publication Format   370
15.4   References   371

Index to Contributors
to Volume 27   373

Preface

Synoptic exploration of the meteorological structure of the upper
atmosphere has proceeded rapidly during the past decade. Two experimental methods are most prominent in this work, one, radar tracking of
ionization tracks of meteors, and two, soundings with physical probes
carried by high altitude research rockets. The Meteorological Rocket
Network (MRN) has provided for synoptic sampling of the stratospheric
circulation on a global basis, and the results obtained from these soundings have demonstrated that the upper atmosphere is meteorologically
active. These results have raised doubts relative to the usual quiescent
neutral upper atmospheric assumptions and necessitate expansion of
synoptic meteorological exploration upward into the region where
electrodynamical processes can transfer energy back and forth between
the neutral and electrical atmospheres. The papers included are focused
on the physical processes which form the structure of the base of the
thermosphere and on the meteor trail radar observational technique for
extending synoptic meteorological exploration into the thermospheric
region.

This book contains selected papers presented at the Summer Institute
titled Physics of the Upper Atmosphere which was conducted by the
Physics Department of the University of Texas at El Paso in cooperation
with the Atmospheric Sciences Laboratory, U. S. Army Electronics
Command, White Sands Missile Range during August 3-14, 1970.

Motivation for synoptic meteorological exploration of the upper
atmosphere came from the needs of rocket test facilities such as White
Sands Missile Range. A detailed understanding is required of the
physical processes involved in determining the atmospheric structural
environment in which rocket testing occurs. The value of any data on
thermospheric structure would be enhanced if it were based on known
atmospheric structure beneath that region, so it is no accident that any
synoptic look at the circulation of the thermosphere would be based on
the MRN synoptic observational stations.

The development of synoptic meteorological exploration by means of
rocketry has been documented since its origin in 1959 in the books,
STRUCTURE OF THE STRATOSPHERE AND MESOSPHERE (W. L.

Webb, Academic Press, Inc., New York, 382 pp., 1966) and STRATO-
SPHERIC CIRCULATION (W. L. Webb, editor, Academic Press, Inc.,
New York, 600 pp., 1969). The first book, dealing with the early days of
synoptic rocketry in America, is Volume 9 in the International Geophysics Series edited by J. Van Mieghem. The second book, which is a
proceedings of the London (1967) and Tokyo (1968) COSPAR (Committee
on Space Research of the International Council of Scientific Unions)
international institutes on the stratospheric circulation, is Volume 22 in
the Progress in Astronautics and Aeronautics Series edited by Martin
Summerfield. This volume presumes a working knowledge of these two
previous volumes and their associated references.

The Atmospheric Sciences Laboratory at White Sands Missile Range
provided the technological impetus for development of the MRN, and
the University of Texas at El Paso provided the educational foundations
for advancement of that effort. At the University this was accomplished
through the Physics Department by special graduate courses in this area
and through the Schellenger Research Laboratory, under contract with
the Atmospheric Sciences Laboratory, by student work on data reduction
and publication in which the first ten thousand MRN rocket soundings
were published in a unified format. In addition to regular graduate courses
in the physics of the upper atmosphere, five natioinal conferences in this
subject area have been hosted by the University and four two-week
summer institutes have been conducted for American MRN personnel,
with these latter efforts forming the basis for American National Academy
of Sciences sponsorship of the London and Tokyo COSPAR institutes.

The 1970 Summer Institute provided a distinguished faculty for laying
the groundwork for further upward expansion of synoptic meteorology.
Dr. C. Sharp Cook, Chairman of the Physics Department of the University
of Texas at El Paso, was the keynote speaker at the dinner on the first
evening of the Summer Institute. He drew on long experience in advanc-
ing technology of the space age in the face of atmospheric constraints,
particularly emphasizing the spectacular interaction of nuclear events
and meteors with the atmosphere and pointing out the urgent need for
communications between the atmospheric sciences community and the
many groups striving to advance modern meteorological technology.
The Summer Institute was designed to produce such communications,
at least within the atmospheric sciences community.

Synoptic exploration of any region of the atmosphere requires a considerable amount of a rather specific kind of data. Systematic variations, such as the annual cycle, tides, etc., are of secondary importance once their structure is established. Essentially the same is true of very small scale features. Since both of these ends of the atmospheric turbulence structure are strong in the upper atmosphere, it is important that the data acquisition, reduction and transmission techniques selected be oriented toward passing that part of the spectrum which contains the maximum amount of synoptic scale information.

This objective has been accomplished in the stratospheric circulation region through use of the Stratospheric Circulation Index (SCI), which consists of the zonal and meridional component wind speeds (mps) in the 45-55 km altitude region. This altitude range was in part dictated by availability of data. In light of the limitations of the Meteor Trail Radar (MTR) wind measuring system, which is considered in this volume, a similar synoptic analysis technique for exploration of the thermospheric circulation woud be a Thermospheric Circulation Index (TCI) which would consist of the mean wind components (mps) in the 90-100 km altitude region. Only a small portion of the available MTR data would be used for synoptic analysis, with the more detailed output reserved for specific studies of eddy diffusion, tides, etc.

Construction of an MTR system at White Sands Missile Range, which has optimum characteristics for synoptic application to exploration of the base of the thermospheric circulation, is under way. Construction of similar radar systems at each of the MRN sites would then give a capability of synoptic exploration of the atmosphere from the surface to 110 km altitude for the first time.

The editor is indebted to Dr. Martin Summerfield, editor-in-chief of the AIAA Progress in Astronautics and Aeronautics Series, and Miss Ruth F. Bryans, AIAA Director of Scientific Publications, for their support in producing this volume, and to Mrs. Glenda McMath for her excellent typing of the final copy.

Willis L. Webb
Editor

June 1971

Foreword

Ancient records carry many references to great meteor showers, meteorite falls and even the motion of persistent luminous trains. Although the fear and mystery have been removed, these phenomena of the interface between our planet and cosmic space can still provide a startling experience for the sophisticated observer who, by very good fortune, is strategically placed at the right time. The detailed physical processes occurring in hypervelocity atmospheric entry of solids from space presents a continuing challenge to various scientific disciplines. The goals of such studies have become surprisingly diversified, including studies of the orbits of the bodies in space, their physical nature, atmospheric densities or motions, environmental conditions, and the processes themselves, sometimes for the sake of ulterior motives, such as re-entry of space vehicles and warlike missiles.

Until the beginning of this century observations were almost completely limited to visual techniques. Thus, the results were largely qualitative, except for the fairly precise determination of radiants and the important conclusion that several meteor streams were definitely the consequence of cometary debris encountering the Earth's atmosphere. The qualitative nature of these observations left wide areas of uncertainty. Poor velocity determinations thus led to the apparently likely possibility that many meteoroids and meteorites pursue hyperbolic orbits about the Sun and therefore are denisons of interstellar space. In 1932 Ernst Opik showed indeed that the Sun could gravitationally control a cloud of comets and meteoric bodies in extremely elongated orbits such that passing stars could occasionally penetrate the cloud. The losses to the system over billions of years he found would be significant but not catastrophic. The passing stars would be analogous to bullets through a cloud of gnats, not endangering the integrity of the cloud as such.

If Opik's hypothesis were indeed true, then we should expect to encounter meteoric bodies gravitationally attached to nearby stars and, of course, some that are freewheeling through space. It is this bold hypothesis that led me into the field of double-station photographic meteor photography in the latter part of the 1930's. Some two decades were required to dispel this attractive hypothesis and still an elusive 1% uncertainty leaves open the possibility of a rare interstellar interloper among our observed meteoric influx. During this period huge improvements were made in the photographic technology, while electronic progress led to radio and radar techniques of great power in determining the ionization and motion of meteorites through the atmosphere. Not only were meteors found to be of solar-systems origin but predominantly cometary debris. Even the large fireballs are of a fragile physical character and probably of cometary origin. A fairly secure foundation of observational data now denies that asteroids passing within the Earth's orbit are all "little Earths"; some perhaps are the dead nuclei of old comets that have become dormant by losing their vitalizing ices. Thus it appears that almost all of the meteors are a consequence of material frozen out of the solar nebula some 4.7 billion years ago.

I find it fascinating that not only the meteors but almost all of the phenomena studied in this important volume are interface phenomena, partly or wholly arising from external radiation or particles from space impinging on our own private space ship, the Earth.

Fred L. Whipple
Director
Smithsonian Astrophysical Observatory

March 16, 1971

CHAPTER 1

UPPER ATMOSPHERIC DYNAMICS

Willis L. Webb

Atmospheric Sciences Laboratory,
U. S. Army Electronics Command,
White Sands Missile Range, New Mexico

1.1   Introduction

The *Thermospheric Circulation* is defined as that geo-circulation system which occupies the atmospheric region above 80 km altitude. The relationship of the thermospheric circulation to the *stratospheric circulation* (25-80 km altitude) and the *tropospheric circulation* (0-25 km altitude) as well as the thermally defined atmospheric boundaries is illustrated in Fig. 1.1. Synoptic meteorological investigations of the tropospheric circulation (see any meteorological text) and the stratospheric circulation (Webb, 1966c,1969) have proceeded to the point that it is now desirable to initiate synoptic exploration of the meteorology of the thermospheric circulation.

The meteorological structure of the the thermospheric circulation region is determined by absorption of hard solar ultraviolet radiation below 2000A (Bates, 1951; Spitzer, 1952; Johnson, 1956, 1958, 1965; Nicolet, 1960; Kallmann, 1961), by conduction of heat earthward from the hot solar atmosphere (Webb, 1970; Johnson and Wilkins, 1966), by dissipation of body waves (Hines, 1960, 1968; Maeda, 1964; Hodges, 1967; Pitteway and Hines, 1963) which have propagated upward from the dense lower atmosphere and by dynamic processes which act to redistribute these inhomogeneous heat inputs. The solar radiant heat input is generally complex (Hinteregger, 1961; Hunt and Van Zandt, 1961), with a portion of the energy first applied to ionization processes which, in the main, is later converted to thermal energy through various recombination processes. Heat conduction down from the hot solar atmosphere is complex because of the intricate interaction between the charged particles of the solar plasma and the geomagnetic field and the highly variable structure of atmospheric eddy diffusion transport coefficients in the base of the thermospheric circulation region.

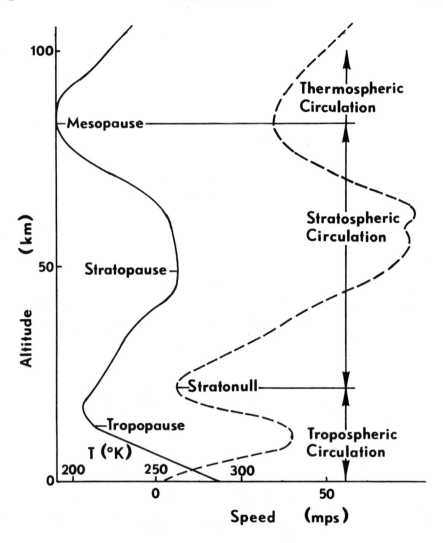

Fig. 1.1 Model temperature (solid curve) and zonal wind (dashed curve) profiles for winter over White Sands Missile Range with the thermal and kinetic nomenclature of vertical subdivisions.

It is relatively new knowledge that a large amount of detailed structure (Webb, 1965; Newell et al., 1966) is characteristic of the neutral atmosphere in the upper regions, and current evidence points strongly toward these detailed features being the result of the presence of internal body waves (Hines, 1960; Jones, 1969) in the rarified gas of the lower thermosphere after that energy has been invested in propagating

modes by lower atmospheric processes. The final mode of establishing lower thermospheric circulation structure is the
one of primary interest here, since any dynamics of the thermospheric circulation forms the heart of meteorological interest
in that region of the atmosphere.

The upper atmosphere has a history of being considered
to be a meteorologically *quiescent* region. Such assumptions
relative to the stratospheric circulation have been shown to
be highly erroneous, principally through the synoptic explora-
tions efforts of the Meteorological Rocket Network (MRN, Webb,
1966 c,1969). New knowledge obtained from the more than 14,000
meteorological rocket soundings of the stratospheric circulation
wind profile have shown that the impact of dynamic processes
on the structure of the upper atmosphere actually increases
with altitude. There is every reason today to expect that synoptic meteorological investigation of the base of the thermospheric circulation will provide important new insights into
the physical processes which control the structure of that
region of the atmosphere.

The electrical structure of the atmosphere becomes of
prominent interest for many technological applications at the
base of the thermosphere. This is basically due to the presence of free electrons which efficiently interact with electromagnetic energy in the radio frequency range and, through
collisions with neutral particles of the lower thermosphere,
severely complicate the electrical characteristics of that
region where detection, tracking and/or communications applications are involved. It can be expected,then, that technological interest in synoptic meteorological knowledge of the
lower thermospheric circulation will center principally on this
problem of interaction between the neutral and electrical atmospheres.

It is of particular interest to note that, while investigations of the electrical structure of the ionosphere over the
past 70 years have centered on assuming that the neutral atmosphere was static and homogeneous, the results of these investigations have invariably indicated the presence of major
dynamic processes through the observed presence of numerous
electrical inhomogeneities. While it has been popular to
conjure physical processes acting from above which could introduce these observed inhomogeneities into a homogeneous
lower atmosphere, there is now adequate reason to suspect that
many, if not most, of the observed inhomogeneities in the
electrical structure are the result of dynamic effects imposed
on the electrical atmosphere from the neutral atmosphere
beneath it.

Observations of the wind field of the thermospheric
circulation region have been conducted sporadically for several
tens of years so that there now exists a limited body of data
which can be used to infer the climatological structure of the
lower portion of the thermospheric circulation. These data
have fallen into two distinct classes; those which are directly
electrical in nature and those which have less obvious electrical characteristics. In the first class are those measurements of the drift of inhomogeneities in ionospheric structure,
either of relatively weak intensity large scale inhomogeneities
or small scale features such as those produced by meteor entries.
The experimental techniques involve observation of electromagnetic energy scattered by the electron components of the
ionospheric gas. The second class of sensors involves observations of the drifts which occur on rocket deployed chemical
trails and on falling inertial particles deployed from rocket
systems. Only in the inertial particles case are wind measurements in the ionosphere clearly free of electrical control.

This is no small point, since the lower thermosphere is
in the heart of the atmospheric region where maximum coupling
exists between the neutral and electrical components of the
atmosphere. This coupling is produced by collisional effects
on charged particle mobility (from the neutral gas point of
view) and by gyro-rotational influences of the geomagnetic
field on charged particle motions (from the electromagnetic
point of view). Collision frequencies and gyro frequencies
of this interaction zone in the 75-150 km altitude region
(see Fig. 1.2), which will here be termed the *dynasphere*,
provide the physical mechanism through which neutral-electrical interaction occurs. The name dynasphere is selected because of the dynamo current systems, a dominant physical feature of this region which are believed to result from interaction between the neutral and electrical atmospheres.

The most efficient observational system for application
to extension of synoptic meteorological dynamic measurements
into the thermospheric circulation region is not at this point
necessarily obvious. This treatise focuses on the meteor trail
radar technique, which was selected for this first attempt
principally because it has certain technical characteristics
which facilitate the application to a synoptic observational
program. These characteristics include the fact that the
sensor is a naturally occurring phenomenon which does not require
large expenditure of effort to deploy, it offers a discrete-
ness which is desirable for these measurements, and it offers
a simplicity in geometry which may lend itself toward the

determination of the actual physics of measurement. The meteor
trail radar technique is severely limited to the altitude range
from 80-110 km. It is not certain at this time that the meteor
trail radar technique offers the best possibility for a synoptic meteorological observational network for the lower thermosphere, but clearly it is a strong candidate for that honor.

There is no rigorous analysis in the literature of the
physics of wind measurement employed in the meteor trail radar
technique in light of our current knowledge of the electrodynamic processes which occur in the dynasphere. The meteor
trail technique is generally considered to produce an acceptable measurement of the neutral winds, based in part on the
fact that the speeds observed are reasonable for upper atmospheric winds, and heavily loaded with assumptions that electrical effects in the dynasphere are negligible. Such assumptions seem questionable when available data indicate that
electric current densities of the order of 10^{-5} amperes per
square meter and electron velocities of the order of kilometers
per second exist in that region of the atmosphere.

The information contained in the following pages then is
an attempt to provide a base from which evaluations of the
sensitivity and accuracy of this observational system can be
made. If the results of such studies should show that the
meteor trail radar technique does indeed provide an acceptable
measure of the neutral atmosphere wind,we will already have in
being an observational system designed to meet our primary
objective. If it should develop that there are electrical
influences on the measurements,it may be possible to separate
those effects and thus observe synoptically both the neutral
and electrical components of the upper atmosphere. In any
case, selection of the meteor trail radar technique for first
application to synoptic meteorological investigation of the
dynasphere can at most be only a small deviation from the
optimum path, since it is not at present obvious that any of
the other acceptable measuring techniques would be less vul-
nerable to error than is the meteor trail radar technique.

1.2 Electrodynamic Processes

Just as the planetary boundary layer provides a special
boundary condition on the base of the tropospheric circulation
in the first kilometer where friction processes are important,
electromagnetic processes of the ionosphere result in development of frictional forces in the upper atmosphere and induce
non-geostrophic conditions into the flow. This meteorological
situation is illustrated in Equation 1.1,

$$\frac{dV}{dt} = 2V \times \boldsymbol{\omega} + \mathbf{g} - \frac{1}{\rho}\boldsymbol{\nabla}p + \mathbf{F} \tag{1.1}$$

where the friction term (**F**) is not generally zero. Other terms in the equation are the velocity **V**, earth's rotation $\boldsymbol{\omega}$, gravity **g**, density ρ, and pressure p.

Frictional effects, regardless of the process from which they derive, have the general effect of altering the wind speed and introducing an imbalance between the pressure gradient and rotational terms of Equation 1.1, with a resulting component of motion between high and low pressure regions. Negative values of the frictional force tend to eliminate meteorological synoptic systems, and positive values tend to intensify those systems, so the magnitudes and signs of these frictional effects are of critical importance to meteorological interests. In general, a fully ionized plasma in a strong magnetic field could have few phenomena of common meteorological interest. In the dilute plasma of the lower thermospheric circulation, however, it is the relative importance of this physical process from which much of the physics of acute meteorological interest must stem.

Frictional effects are generally induced through collisional processes. These collisional processes are largely mechanical in the surface boundary layer, with momentum exchange occurring between individual air molecules and the rigid surface providing the reduction in wind speed which produces the imbalance between rotational and pressure gradient forces (Equation 1.1). More complex interactions become important when the flow is turbulent, and various state changes serve to induce these turbulent viscous effects into deeper layers of the lower atmosphere. In the ionosphere, the frictional processes are also collisional in nature, with the interaction occurring between neutral molecules which tend to move in response to the total pressure gradient and charged molecules which are loosely tied to the geomagnetic field through electromagnetic effects.

Motions of charged particles in the atmosphere are controlled by the processes indicated by the equilibrium condition illustrated in Equation 1.2,

$$q(\mathbf{E} + \mathbf{v} \times \mathbf{B}) + M\mathbf{g} + M\nu\mathbf{w} - \frac{1}{n}\boldsymbol{\nabla}p = 0 \tag{1.2}$$

where q is the electrical charge of the particle, **E** is the electric field, **v** is the motion of the charged particle normal to the magnetic field **B**, M is the mass of the charged particle, **g** is gravity, **w** is the speed of the charged particle through the neutral gas, ν is the collision frequency (Fig. 1.2),

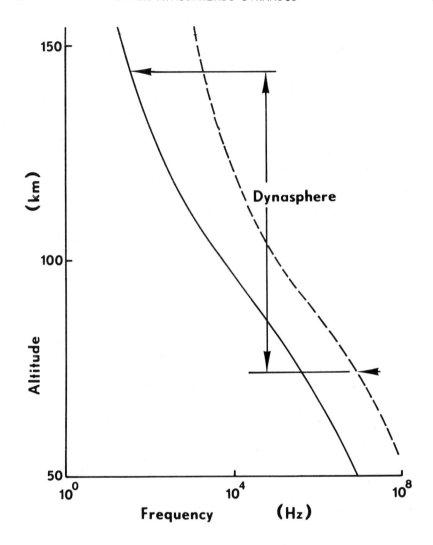

Fig. 1.2 Collision frequencies for ions (O_2, solid curve) and electrons (dashed curve) with neutral molecules and their equatorial gyrofrequencies (arrows).

n is the density and ∇p is the pressure gradient of the charged particles.

An important electromagnetic parameter involved in neutral-electrical interactions is the *gyrofrequency* (ω_g) of these charged particles which is given by the relation presented in Equation 1.3,

$$\omega_g = \frac{qB}{M} \qquad (1.3)$$

Matched with these detailed geomagnetic motions are the
collision frequencies between charged and neutral particles,
representative model vertical profiles of the more important
of which are presented in Fig. 1.2. Gyrofrequencies of electrons and molecular oxygen ions in equatorial regions of the
dynasphere are illustrated by the arrows, with neutral pressure gradient forces in control of the charged particle motions
when the collision frequencies are greater than the gyrofrequencies (below approximately 75 km for electrons and 150 km
for ions) and electromagnetic forces tending to control motions
of the charged particles above those levels.

The gyro motion forces (Equation 1.2) on charged particles
in the dynasphere by the geomagnetic field do not alter the
motion in the direction of the magnetic field, but do introduce entirely new processes relative to motions normal to the
magnetic field. The *electrical conductivity* along the magnetic
field is then given by

$$\sigma = \frac{nq}{B} \left[\frac{\omega_i}{\nu_i} + \frac{\omega_e}{\nu_e} \right] \tag{1.4}$$

where the subscripts i and e indicate ions and electrons, respectively. This relation illustrates the usual static concept of collisions reducing the conductivity in the direction
of this *specific* conductivity.

An ambient electric field normal to a magnetic field
introduces asymmetries in the motions of charged particles.
If collisional frequencies are small relative to the gyro
frequencies (above 75 km for electrons and 150 km for ions)
the conductivities in directions normal to the magnetic field
will be greatly reduced, with the maximum reduction in the
direction of the electric field. The conductivity in this
Pederson case is given by

$$\sigma' = \frac{nq}{B} \left[\frac{\nu_i \omega_i}{\nu_i^2 + \omega_i^2} + \frac{\nu_e \omega_e}{\nu_e^2 + \omega_e^2} \right] \tag{1.5}$$

Normal to the plane of the magnetic and electric vectors
the conductivity assumes an intermediate value which is termed
the *Hall* conductivity. The magnitude of the Hall conductivity
is given by

$$\sigma'' = \frac{nq}{B} \left[\frac{\omega_i^2}{\nu_i^2 + \omega_i^2} - \frac{\omega_e^2}{\nu_e^2 + \omega_e^2} \right] \tag{1.6}$$

Using nominal values of the several atmospheric variables
model conductivity profiles for the specific, Pederson and
Hall directions at noontime in low latitudes are illustrated in
Figure 1.3. These curves indicate that the lower ionosphere

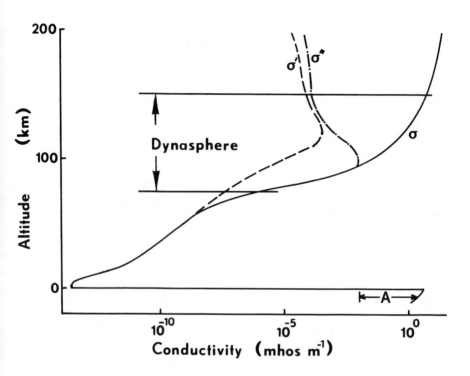

Fig. 1.3 Model noontime low latitude direct current specific (solid curve), Pederson (dashed curve) and Hall (dash-dot curve) electrical conductivity structures of the atmosphere and the earth's surface (A).

is a rather good conductor of electricity, with values somewhat less at 100 km altitude than that for seawater in the specific conductivity case, but reduced more than three orders of magnitude in the Pederson direction and less than an order of magnitude in the Hall direction at that same altitude. In the earth's atmospheric system the specific conductivity is oriented roughly in the meridional direction, the Pederson conductivity is roughly vertical (since that is the only direction in which a charge separation can be effectively maintained) and the Hall conductivity is roughly zonal.

A most interesting aspect of plasma physics is contained in the *Hall effect*. Consideration of the dynamics of motion of positively and negatively charged particles in the presence of perpendicular magnetic and electric fields shows that they deviate in a very special way from the nice symmetry of the circular (in opposite directions) motions (Equation 1.3) exhibited when only a magnetic field is present. As is illus-

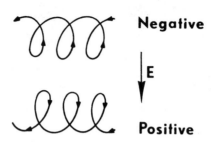

Negative

E

Positive

Fig. 1.4 Motions of equal mass positive and negative particles in normal perpendicular magnetic and electric fields.
The magnetic field is pointing out of the paper while the
electric field has the direction indicated by the error.

trated in Fig. 1.4, the accelerating and decelerating forces
imposed on the charged particle's gyromotion by the electric
field result in a migration of both sign particles in the
same direction. This mean motion is summed up in Equation 1.7,

$$U = \frac{E \times B}{|B|^2} \qquad (1.7)$$

which is the *Hall drift velocity* imparted to charged particles
(without regard to sign) in the absence of collisions. Now
in the dynasphere the electrons can,in part,execute these
motions because they are relatively free of collisions, but ions
cannot take part in the Hall drift since collisions with neutral particles will prevent their executing the gyro-rotation.
The net result of this gross difference in electromagnetic
processes is that an electric current will be produced and/or
charge will be separated; i. e., the process will serve as an
electromotive force.

The high conductivity of the ionosphere (Fig. 1.3) will
provide for low impedance relaxation current paths for any
such horizontally oriented electromotive force so that the net
effect will be principally dynamic, with charge separations
and potential variations kept at relatively low levels. In
the vertical, however, it is clear that larger *charge separations* can be impressed on the ionosphere since there is a very
poor return current path for any motions of charged particles
imposed on the lower ionosphere in the vertical direction.
The maximum opportunity for exchange of energy between the

neutral and electrical atmospheres is then to be expected
when vertical motions and electric fields are involved.

The equilibrium force equation for a charged particle
can be approximated for ions by equating the first and fourth
terms of Equation 1.2 as is indicated in Equation 1.8,

$$q\,\mathbf{E} = M\nu\mathbf{w} \qquad\qquad (1.8)$$

The mass of the subject charged particle, its collision frequency with the ambient neutral particles and the speed of the
neutral particles across the magnetic field then serve to establish an opposing static electric field which will prevent
further charge separation if one of the groups of oppositely
charged particles is, for any reason, prevented from moving
in the direction of the wind. Effectively, under equilibrium
conditions, electrical forces will force the collision-controlled
charged particles to 'swim' upstream at a speed equal to the
wind speed.

The magnitude of the negative frictional force (Equation
1.1) which will be produced by the neutral (n_n) atmosphere
transporting a weak plasma (n_i) across a strong magnetic
field (I- magnetic dip angle) can be estimated from the
relation

$$F = \frac{\nu_i n_i u}{n_n}\ \sin^2 I. \qquad\qquad (1.9)$$

Substitution of reasonable values for the various atmospheric
parameters indicates that this electromagnetic force on the
neutral motions becomes of importance in the total balance of
forces in the upper portions of the dynasphere and will surely
be the dominant process for establishing equilibrium conditions
in the upper atmosphere.

Above that altitude at which these electrodynamic forces
become greater than the rotational forces, the equilibrium condition of Equation 1.1 can be approximated for the zonal case
by the expression

$$\frac{\nu_i n_i u}{n_n}\ \sin^2 I = -\ \frac{1}{\rho}\ \frac{\partial p}{\partial y} \qquad\qquad (1.10)$$

where the pressure gradient force is now balanced by the charged
particle drag. Since this force is significant only for zonal
and vertical components of the wind at middle and low latitudes,
it is clear that meridional motions must be of considerable
significance in the upper atmosphere structure.

All of the above considerations have been formulated under

the constraints of a direct current system; i. e., the frequency of any variations is very low. While this model can
reasonably be used for seasonal and climatological type studies
of neutral-electrical interactions, it is obvious that short
period variations such as those naturally produced by gravity
waves and turbulence and by man-made perturbations of a variety
of types will not be satisfied by such restricted conditions.
Equations 1.4, 1.5 and 1.6 must then be rewritten (using
Equation 1.3) to include the frequency (f) of the impressed
signal

$$\sigma = nq^2 \left[\frac{1}{M_e(\nu_e-if)} + \frac{1}{M_i(\nu_i-if)} \right] \tag{1.11}$$

$$\sigma' = nq^2 \left[\frac{\nu_e-if}{M_e\{(\nu_e-if)^2+\omega_e^2\}} + \frac{\nu_e-if}{M_i\{(\nu_i-if)^2+\omega_i^2\}} \right] \tag{1.12}$$

$$\sigma'' = nq^2 \left[-\frac{\omega_e}{M_e\{(\nu_e-if)^2+\omega_e^2\}} + \frac{\omega_i}{M_i\{(\nu_i-if)^2+\omega_i^2\}} \right] \tag{1.13}$$

The variations in conductivity indicated by these relations,
which are complex in structure, not only will introduce inhomogeneities into the propagation of any signal, but also will
introduce nonlinear effects as a result of extraordinary propagation processes.

1.3 Neutral Upper Atmospheric Dynamics

Synoptic exploration of the stratospheric circulation
(25-80 km) has revealed a *monsoonal circulation* system (Webb,
1964) which is well illustrated by the annual SCI curves presented in Fig. 1.5. These data show a marked asymmetry between the summer and winter seasons, both in duration and in
structure. These facts have been interpreted as indicating
that westerlies dominate the global stratopause region at
equinox time and easterlies dominate that region at solstice
times. While the obvious features of solar heating can be re-
conciled with the equinox situation, it is only with the addition of strong vertical eddy transport as well as global circulation systems that reasonable models of the solstice time
stratospheric circulation structure can be obtained.

The best examples of these deviations away from static
structure are the low to middle latitude high pressure belt of
winter which marks separation of the summer easterlies and
winter westerlies in the stratopause region and the cold mesopause (near 80 km altitude) of summer high latitudes (Webb,
1966) and warm mesopause of winter high latitudes. The meso-

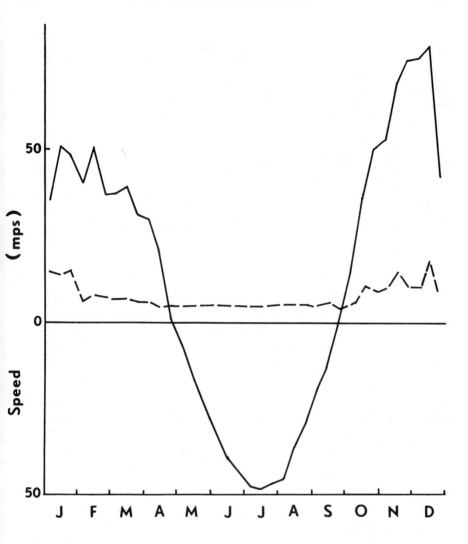

Fig. 1.5 Zonal (solid curve) and meridional (dashed curve) SCI ten day mean values for all data obtained at White Sands Missile Range through 29 June 1970.

pause case is of great importance to our objective of synoptically exploring the thermosphere. Detailed inspection of this solstice mesopause structure has shown that the coldest temperatures of the atmosphere are found in a ring around the summer pole near 60° latitude (Morris and Kays, 1969), with the maximum summer easterly winds occurring in the stratospheric circulation in polar regions at mesopause altitudes.

Equally impressive is the fact that the stratopause is observed to contain an asymmetric diurnal *thermal tidal variation* of the order of 15 °C temperature range and 30 mps wind variation (Beyers and Miers, 1965; Miers, 1965). This variation is several times *larger* than had been estimated from theoretical considerations based on static assumptions (Leovy, 1964; Lindzen, 1967), and these variations cannot be considered small relative to the mean quantities of the region. The overall global structure of the stratospheric circulation thermal tides, based on observational data, is illustrated by the solstice time situation presented in Fig. 1.6 (Webb, 1966a, 1967). It is important to note that the minimum altitude of the base of the tides (near 50 km) will be found near the subsolar point, but that the altitude of these phenomena will be higher near the sunrise-sunset circle, particularly at high latitudes where equilibrium can more nearly be achieved.

A third important result which has been attained from small meteorological rocket exploration of the stratospheric circulation is the fact that the upper atmosphere is loaded with *small scale structure* (Webb, 1965; Newell et al., 1966). The vertical sizes of these features are obvious in all sensitive measurements of the vertical wind, temperature and ozone concentration profiles. A small amount of these data has been analyzed to obtain the mean structure illustrated in Fig. 1.7. These data are in general agreement with vertical dimensions of small scale features of the troposphere and are in excellent agreement with those data obtained by Greenhow and Neufeld (1959) through detailed analysis of meteor trail wind profiles in the 90 km altitude region. These large values of eddy transport have been shown to have gross impacts on the thermal (Webb, 1969), chemical (Hesstvedt, 1968) and electrical (Webb, 1968) structures of the atmosphere.

Both molecular and eddy transports serve to produce a flux (F) of minor constituents down any gradient and thus result in a trend toward uniformity in accord with the relation illustrated in Equation 1.14,

$$F = -K \frac{\partial n}{\partial x} \tag{1.14}$$

where n is the concentration and x is normal to contours of equal concentration. In the thermal case of vertical heat diffusion the situation is quite different, with molecular diffusion transport (F_m) following the similar relation

$$F_m = - K_m \rho c_p \frac{\partial T}{\partial h} \tag{1.15}$$

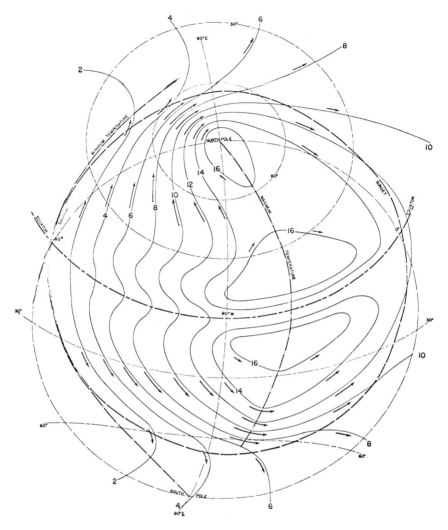

Fig. 1.6 Model temperature (°C) and wind field of the stratopause heat wave at summer solstice in the Northern Hemisphere
(Courtesy of the North-Holland Publishing Company).

where c_p is the specific heat at constant pressure. Eddy heat
transport (F_e) in the vertical, on the other hand, forces the
temperature structure to assume the configuration indicated by
Equation 1.7

$$F_e = - K_e \rho c_p \left(\frac{\partial T}{\partial h} + \gamma\right) \qquad (1.16)$$

where γ is the adiabatic lapse rate of 10°C per km. At some
altitudes the values of eddy transport indicated in Fig. 1.6
produce downward heat fluxes which are greater than the total

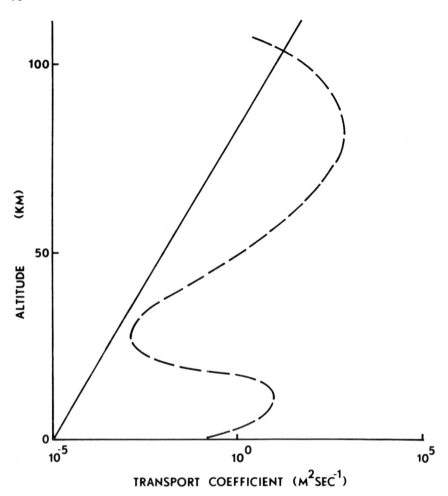

Fig. 1.7 Model profiles of atmospheric molecular (solid curve) and eddy (dashed curve) transport coefficients based on Lettau (1951) in the neutral atmosphere and Booker (1956) in the ionosphere.

estimated heat input from the sun above that level (Johnson and Wilkens, 1965). This can be interpreted as showing that where such large values of eddy transport exist there must be additional heat sources derived from dynamic (adiabatic heating), chemical and/or electrical heating effects. In any case, it is clear that solar insolation heating of the atmosphere falls far short of being an adequate approximation to the actual situation, and those cases of unusual thermal structure (such as the cold mesopause in summer high latitudes) must have dynamic contributions.

1.4 Neutral-Electrical Interactions

In the final analysis any understanding of *coupling* between the neutral and electrical components of the earth's atmosphere will rest on relevant observational data. As was indicated by the very brief survey of the results of the first decade of synoptic rocket investigation of the stratospheric circulation presented in Section 1.3, these new data point clearly to the fact that the neutral upper atmosphere is characterized by a complex spectrum of dynamic processes. Thus, the observed complexities of ionospheric structure no longer appear *strange* and the quest for the origin of the observed variability need not look out toward unknown sources, but rather can be directed at neglected facets of dynamic interactions between these intimately related atmospheric components.

Investigation of neutral-electrical coupling is not a new field, as is evidenced by the summaries by Maeda and Maeda (1969) Lauter, Sprenger and Entzian (1969) and Sechrist of this volume (Chapter 11). The problem has centered on lack of proper data, particularly relative to neutral atmosphere motions in ionospheric regions. At most, a very small sample has been available and the strong belief that synoptic processes would become negligible in the rare upper atmosphere has served to reduce the assumed role of neutral forcing of ionospheric structure to very modest proportions. The synoptic view of the stratospheric circulation provided by the MRN has effectively reversed this minimizing concept, and now it is time to inspect 'in situ' the dynamic structures of both of these important components of the atmosphere.

The meteor trail radar offers an opportunity to obtain, on a synoptic basis, the data required to probe upper atmospheric neutral dynamics for the physical processes which control the structure of the dynasphere and facilitates transfer of signals from one atmospheric component to the other. However, current knowledge of the complexity of upper atmospheric structure already warns us that, even with a meteor trail radar at each MRN station, it would still be necessary for some time to depend on other more general sources of knowledge of atmospheric structure to set boundary conditions on the processes which are considered. As always in science, observations provide the guide if we are able to understand the indicators.

It has long been known that the fair weather electric field of the atmosphere near the earth's surface (approximately 100 vm^{-1}) exhibits, when measurements are made away from local disturbing influences, a *universal time diurnal variation* with

a maximum near 1800Z (Mauchly, 1923) as is illustrated in the
upper curve of Fig. 1.8.   This electromotive force drives a

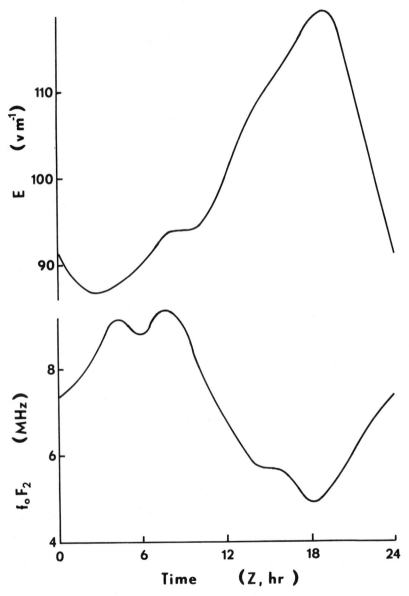

Fig. 1.8 Diurnal variation with universal time of the oceanic
fair weather electric potential gradient (upper curve) observed
by Mauchly (1923) and F region critical frequency (inversely related to electron density) of Antarctica reported by Duncan
(1962).

current density of roughly 10^{-12} am^{-2} toward the earth which,
over the fair weather portion of the earth, adds up to a total
tropospheric electric current of approximately 1500 amperes
(Chalmers, 1957; Israel, 1957). The hypothesis that thunderstorms provide the return current path (as well as the electromotive force) for this tropospheric electrical system led to
investigation of their diurnal variations, with the resulting
find that the 1900LT maximum observed generally does indeed
exhibit an amplitude variation with location with a peak near
1800Z (Appleton, 1925; Whipple, 1929; Brooks, 1925).

Not so well known is the fact that Duncan (1962), following the lead of Bellchambers and Piggott (1958) and Coroniti
and Penndorf (1959), has reported a comparable universal time
variation in the electron density of the F region (near 300 km)
of the ionosphere in Antarctica. A summary of these data,
presented in the lower curve of Fig. 1.8 in terms of the
critical frequency of reflection, provides a most interesting
comparison between these very different electrical environments.
The reader should note that the morphology of these data is
not without controversy (King et al., 1968; Duncan, 1969;
Challinor, 1970), with alternate reasons postulated for the
variable situation which is observed at other locations. It
is also interesting in this regard to note that the fair weather
electric potential gradient curve is not representative of the
universal time diurnal variation in areas where local sources
of electrical perturbation structure dominate. In a complex
environment such as the earth's atmosphere it is not possible
with these limited data to be sure that the interesting relations to be inferred from these curves are even remotely in
response to each other.

These data do show, whatever the processes involved, that
the electron density of the F region evidences a maximum at the
same location on the earth at which the tropospheric (and
thunderstorm frequency) current system is a maximum; namely,
on the 1900Z meridian.

Looking to available data located closer to the suspected
seat of neutral-electrical interaction it is interesting to
compare the details of the observed stratospheric circulation
which operates up to the base of the dynasphere (80 km). A
most important global scale perturbation of this region is the
thermal tidal circulation systems (Fig. 1.7) generated by ozone
absorption, with a peak intensity near the subsolar point near
50 km altitude (Beyers and Miers, 1965; Miers, 1965; Webb, 1966).
MRN observational data have indicated that the easterly segment
of the summer hemisphere thermal tidal circulation (the Stratospheric Tidal Jet; Webb, 1966) is fundamental to production of

the coldest temperatures in the atmosphere at the mesopause
near 60° latitude, where easterly winds near 100 mps occur with
the very important (dynamically) noctilucent clouds (see Chap-
ters 5 & 6).

The events related to shift of the tidal circulation system from a westerly segment short of the polar region during
fall equinox, winter and the spring equinox, period to the
summer time mode of encircling the summer polar region are
indicated in the mean SCI data for Fort Greely as is illustrated in Fig. 1.9. These data, along with the noctilucent
cloud occurrence data presented in the lower curve of Fig. 1.9,
led to inspection of the high latitude summer mesopause region
for the stratospheric tidal jet.

Since the tidal circulations can be expected to produce
electrical effects in the dynasphere (Webb, 1968, 1969, 1970),
and since cloudy condensation can always be expected to materi-
ally alter the local electrical structure, the search for interaction inferences was focused on the most obvious electrical
phenomenon in that region, the dynamo currents. The Sq variations of the geomagnetic field imposed on the earth's permanent
magnetic field have long been studied (Chapman and Bartels,
1941; Matsushita, 1965). One of the interesting results of this
work was reported by Weise (1951), and again by Wagner (1968,
1969) as a *semi-annual* variation in the intensity of the diurnal
Sq variation.

Simple inspection of their data, as is illustrated in the
middle curve of Fig. 1.9, indicates that the perturbations
which they detected in the annual variation of the Sq system
are well represented by a semiannual cyclic variation, but
rather appears to be a special reduction in summertime Sq
intensity associated with development and decay of the summer
season circulation. Of special interest here is the special
temporal lineup of these events in the stratopause region
(∼50 km), the noctilucent cloud zone (∼80 km) and the dynamo
current region (∼100 km). The clouds start to form and the
dynamos weaken as the tides shift to encircle the summer polar
region, while the clouds abruptly stop and the dynamos pick up
again as the summer easterlies begin their precipitous decay.
These data provide strong inference that the physical processes
which shape the structure of the neutral upper atmosphere also
exert a controlling influence on the electrical structure of the
base of the ionosphere.

Ionospheric scientists have studied motions of inhomogeneities in electron density structure of the lower ionosphere
(under the assumption that they provide a measure of *neutral*

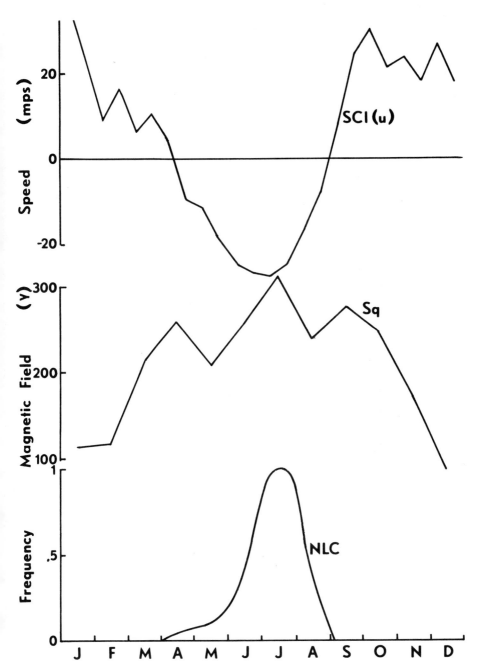

Fig. 1.9   Zonal SCI (10 day means) for Fort Greely during the period 6 November 1963 to 29 June 1970 (upper curve), Sq 'summed ranges' after Wagner (1968, 1969) (middle curve) and North American noctilucent cloud occurrences after Fogle (1968).

winds with special radar systems. An example of the type of
data they obtain in equatorial regions is well illustrated in
Fig. 1.10 by the curve (dashed) reported by Chandra and Rastogi
(1969) for the Thumba dynamo current region (near 100 km altitude). They observe motions from the east throughout the year
in the 100-200 mps range, with the annual variation of monthly
means presented in Fig. 1.10.

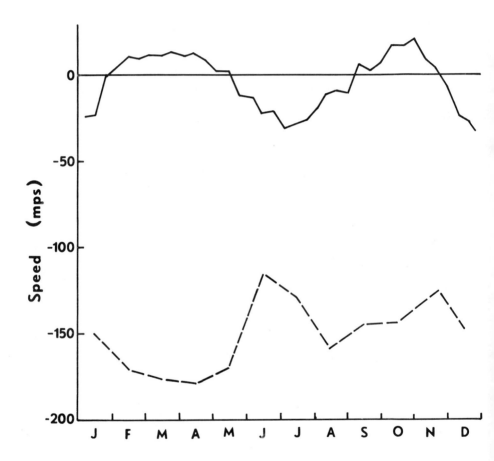

Fig. 1.10 Equatorial zonal SCI illustrated by 10 day means
observed at Panama (9°20'N, 79°59'W) over the period 3 September 1966 to 3 April 1970 (solid curve) and drift of E region
inhomogeneities (Chandra and Rastogi, 1969) at Thumbe (8°33'N,
76°52'E; 0.3°S geomagnetic) over the period 1964-1967 (dashed
curve).

These data do not look too strange to meteorologists of
the MRN since they are well experienced with semiannual

variations in the circulation of tropical regions (Webb, 1964).
Ten-day means of the zonal winds observed over Panama are illustrated by the SCI data of the upper curve of Fig. 1.10, which
points out the fact that in equatorial regions at 50 km altitude
westerly winds dominate at equinox times and easterly winds
dominate at solstice times, resulting in a well defined (although
not symmetric) semiannual variation.

The inverse relationship indicated here is interesting,
with the easterly motions indicated by E region drifts increasing when west winds are present in the stratospheric circulation and decreasing when easterly winds are present below the
E region. It is hazardous to guess about tropical circulation
systems, as is evidenced by the unexpected discovery of the
biennial cycle at 25 km altitude over the rotational equator
(Reed, 1962), but it is obvious that meteor trail radar measurements in equatorial regions would be a most intelligent first
move in attempting to study neutral-electrical interactions in
the upper atmosphere.

Perhaps it is fitting that this short listing of data
which indicate coordinated action of a total complex atmosphere should include an illustration which is of more apparent
synoptic meteorological interest. The data presented in
Fig. 1.11 were obtained at White Sands Missile Range during the
period 23 December 1970 through 3 January 1971. The upper
curve indicates the difference in reading between two rubidium
vapor magnetometers (courtesy of James Johnson of the Atmospheric Sciences Laboratory) after an average difference of
roughly 3300γ (which results because they are located 983 meters
apart in altitude) has been removed. The normal diurnal difference between their Sq amplitudes is approximately 1γ, with the
upper instrument generally reading higher. The very striking
event recorded here can only be appreciated after inspection
of a year's record (these data have not been published).

Irrespective of the cause, a strong perturbation appears to
have occurred in the *dynamo currents* over White Sands Missile
Range, starting on Christmas Day and lasting until 29 December.
This event occurred in an atmosphere which was relatively
quiescent (maximum surface temperatures near 60°F) for the winter season, with no major weather systems in evidence in the
area during the latter part of December. The storm broke at
the surface on 1 January, with the lowest pressure (lower curve
of Fig. 1.11) of record observed late on 2 January, with strong
winds (maximum 49 kts) and minor snowfalls accompanying an influx of cold air which produced minimum temperatures of 9°F.

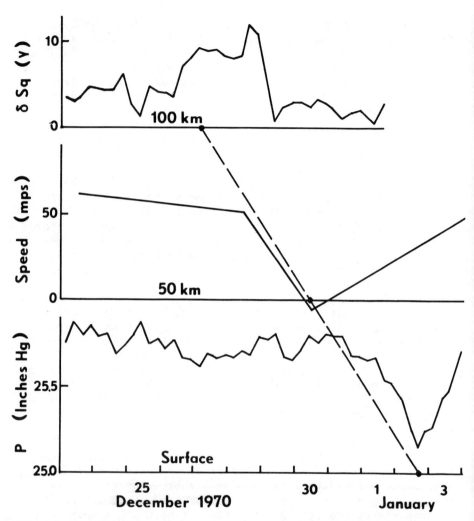

Fig. 1.11   Observations at White Sands Missile Range of deviations from normal (upper curve) of two rubidium vapor
magnetometer differences assumed to result from synoptically
induced variations in the dynamo currents (δSq) near 100 km,
the zonal SCI at 50 km (middle curve) and the surface pressure
(lower curve).

Four meteorological rocket soundings were obtained during
the period with a marked variation in the zonal wind as is
illustrated by the middle curve of Fig. 1.11.   Westerly winds
in the 50-60 mps speed range are indicated for the fine weather
period during which the dynamo current perturbation began.
These speeds are somewhat higher than average for the winter
storm period, which was heralded by an initial decrease by a

factor of two of the 50 km winds over White Sands Missile Range
during early December. The passage of a rather local circulation system at the stratopause level is indicated by the light
easterly winds observed on 30 December, with a return to roughly
normal winter westerlies by early January.

There is contained in these events the inference that they
are not independent, but that they result from a comprehensive
perturbation of the entire atmospheric system. If such a hypothesis should prove to be true, the sequence of events is
illustrated by the dashed line, starting at the 100 km level
and working down to the surface. Clearly, data from one station
will never be adequate to establish such continuity, and it is
only with synoptic observational systems operating in each of
these regions (preferably continuously with altitude) that the
dynamic processes producing these variations will be identified.

Inspection of relationships between the intensity of the
stratospheric circulation and the tropospheric circulation
systems for the winter storm periods of record has provided
evidence that similar decreases in the stratospheric zonal
circulation generally precede by one to three days strong
circulation perturbations in the tropospheric circulation. Extension of this continuity to the dynamo region is one of the
more important aspects of synoptic exploration of the dynaspheric region of the thermospheric circulation.

1.5 Dynaspheric Observational Program

Extensive observations of meteor trail motions have been
made at Jodrell Bank (53°N, 2°W) and at Adelaide (35°S, 138°E).
These data have been reported by Greenhow and Neufeld (1961)
and Elford (1959) and have been summarized by Kochanski (1963).
Similar data for the Soviet Union area have been summarized by
Lysenko (1963). Recently Hook (1970) has summarized two years
of meteor trail data obtained at College, Alaska (65°N, 149°W).
In addition, smoke trail observations of dynaspheric winds at
Wallops Island (38°N, 75.5°W) have been summarized by Kochanski
(1964), at Eglin (30°N, 87°W) by Rosenberg and Justus (1966)
and at Barbados (13.1°N, 57.5°W) by Murphy, Bull and Edwards
(1966).

The nature of the Jodrell Bank and Adelaide circulation
data is well illustrated by the zonal means presented in Fig.
1.12 and the meridional means presented in Fig. 1.13. Most
significant in the data from these two stations is the marked
difference in annual character of the zonal winds. The Adelaide
data (dash-dot curve) indicate strong westerly winds with a
weak period of easterly winds around the September equinox.

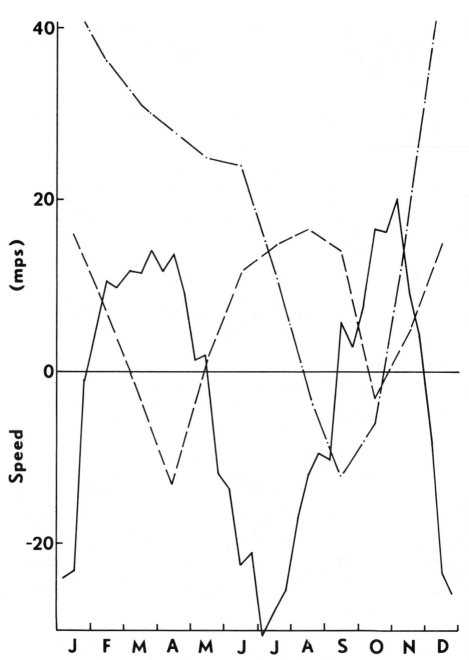

Fig. 1.12 Mean zonal SCI (50 km) for Panama (9°N, solid curve)
and dynaspheric means for Jodrell Bank (53°N, dashed curve) and
Adelaide (35°S, dash-dot curve). The Adelaide curve was obtained by simple averaging of the three curves (80, 92 and 100 km)
published by Elfors (1959).

Fig. 1.13 Mean meridional SCI (50 km) for Panama, 9°N (solid curve) and dynaspheric means for Jodrell Bank, 53°N (dashed curve) and Adelaide, 35°S (dash-dot curve). The Adelaide curve was obtained by simple averaging of the three curves (80, 92 and 100 km) published by Elford (1959).

Jodrell Bank, on the other hand, exhibits (dashed curve) weak westerly winds near solstice times and weaker easterly winds near equinox times. Easterly winds in the April minimum are significantly stronger than are the easterlies of the October minimum.

The dynaspheric circulation of Southern Hemisphere low latitudes appears, from these data, to be greatly different from that observed in Northern Hemisphere high latitudes. This gross difference could result from latitudinal variations in the zonal winds or it could result from hemispheric differences in the dynaspheric circulation. Experience with the synoptic data of the MRN in the stratospheric circulation case favors the former hypothesis, since those data indicates that latitudinal variations are a general characteristic of the upper atmosphere and that the hemispheric circulations, while there are differences, have generally similar characteristics.

For reference purposes monthly mean zonal winds observed

at Panama are included in Fig. 1.12. These data indicate that
easterly winds occurring near the high latitude 100 km level are
matched with westerlies in equatorial regions near 50 km. The
dynaspheric circulation is from the west near solstice times
while the tropical stratopause region has easterly winds. The
semiannual variation of stratopause zonal winds has already been
identified (Webb, 1964) with hemispheric interaction processes
in the stratospheric circulation, and even this limited amount
of data raises interesting points relative to the relationship
between these circulation systems.

If the Adelaide data may be considered representative of
the dynaspheric circulation at low latitudes, they indicate
another major change in equatorial circulation with height.
From the strong biennial variation at 25 km altitude the circulation shifts to a strong semiannual variation at 50 km and
then, apparently, to a solid annual variation at 100 km. Synoptic data on a global scale throughout the upper atmosphere
will be required before the circulation systems indicated here
can be clearly described. In this regard, the data for Obninsk
(55°15'N, 36°41'E) and Kharkov (55°45'N, 37°45'E) published
by Kashcheyev and Lysenko (1968) indicate that the semiannual
variation observed at Jodrell Bank is a general feature of the
dynaspheric circulation in that region.

The meridional data for Jodrell Bank and Adelaide for the
dynaspheric region presented in Fig. 1.13 illustrate the most
interesting fact that there is a global wind from summer polar
regions to winter polar regions indicated by these data. Such
a meridional flow had already been hypothesized on the basis of
the need for subsidence in winter high latitudes to maintain
the observed warm mesopause temperatures and the need for adiabatic cooling at summer high latitudes to produce the atmosphere's lowest temperatures. Inspection of the Obninsk and
Kharkov meridional data indicates this same type of circulation
at those locations. This is in marked contrast to the Panama
(solid curve) meridional data, which exhibits a steady equator-
to-pole flow for the 50 km region because of a diurnal bias
in the MRN observational time which dominantly measures the
thermal tides as they diverge from the equatorial region.

These data then indicate that the dynaspheric portion of
the thermospheric circulation system must be synoptically explored before the meteorological structure of the upper atmosphere
can be derived. This synoptic exploration can be efficiently
done with meteor trail radar systems located at MRN stations.
At these locations vertical wind profiles will then be available from the surface to 110 km, avoiding the pitfalls which
invariably attend voids in any set of data. The meteor trail

radar system at White Sands Missile Range (see Chapter 14) was
designed to efficiently provide such atmospheric data at this
rocket test location, and it serves well as a pattern for systems which can be used at other MRN stations.

Meteor trail radars have been used for a number of years
at various locations over the globe. Most of them were built
and operated for reasons other than meteorological, so in many
cases they were overdesigned for meteorological observational
purposes, and in most cases the data they produced were simply
not reduced into wind information. The sites of known meteor
trail radar sets which are capable of producing wind data are
indicated in Fig. 1.14. It is clear that a strong base already
exists for establishment of a dynaspheric synoptic observational
network with meteor trail radars as the measuring system.

There is today no organized flow of data on the synoptic
aspects of the neutral dynamics of the dynaspheric region. The
most important aspect of such an effort centers in getting the
data in a useful form for research scientists to use in analysis of synoptic scale meteorological systems. As is outlined
in Chapter 15, a TCI (Thermospheric Circulation Index) would
provide for immediate coordination of synoptic information on
the meteorology of the dynaspheric region, leaving most of
the data collected by an MTR station for more detailed inspection in special research projects.

This volume is designed to bring to research scientists
the information required to address actively the problems
which are associated with synoptic exploration of the dynaspheric region. There can be little doubt that such a cooperative effort will produce scientific results which will more
than justify the required expenditures. A bibliography of
many of the past papers dealing with meteor trail wind measurements and associated phenomena is included for the guidance
of the serious student of this area of the atmospheric sciences.

1.6 References

Appleton, E. V., 1925. "Discussion on Ionization in the Atmosphere." Proceedings of the Physical Society of London, Vol.
37, 48-50D.

Bates, D. R., 1951. "The Temperature of the Upper Atmosphere."
Proceedings of the Physical Society of London, B, Vol. 64,
805-821.

Bellchambers, W. H., and W. R. Piggott, 1958. "Ionospheric
Measurements made at Halley Bay." Nature, Vol. 182, 1596-1597.

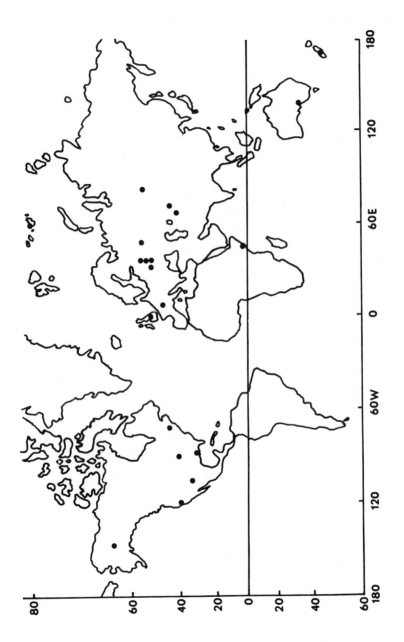

Fig. 1.14 Locations of Meteor Trail Radar stations which are designed to produce wind data from meteor trail drifts.

Beyers, N. J., and B. T. Miers, 1965. "Diurnal Temperature Change in the Atmosphere Between 30 and 60 km over White Sands Missile Range." Journal of Atmospheric Sciences, Vol. 22, No. 3, 262-266.

Booker, H. G., 1965. "Turbulence in the Ionosphere With Application to Meteor Trails, Radio Star Scintillations, Auroral Radar Echoes, and Other Phenomena." Journal of Geophysical Research, Vol. 61, 673-705.

Brooks, C. E. P., 1925. "The Distribution of Thunderstorms over the Globe." Geophysics Memoir, London, Vol. 24, 264-267.

Challinor, R. A., 1970. "The Behavior of the Arctic F-Region in Winter." Journal of Atmospheric and Terrestrial Physics, Vol. 32, 1959-1965.

Chalmers, J. A., 1957. Atmospheric Electricity, Pergamon Press, Oxford, 327 p.

Chandra, H., and R. G. Rastogi, 1969. "Horizontal Drifts in the E- and F-Regions over Thumba During Day-Time." Journal of Atmospheric and Terrestrial Physics, Vol. 31, 1205-1215.

Chapman, S., and J. Bartels, 1941. Geomagnetism, Volumes 1 & 2, Clarendon Press, Oxford.

Coroniti, S. C., and R. Penndorf, 1959. "The Diurnal and Annual Variations of f_oF_2 over the Polar Regions." Journal of Geophysical Research, Vol. 64, 5-18.

Duncan, R. A., 1962. "Universal Time Control of the Arctic and Antarctic F Region." Journal of Geophysical Research, Vol. 67, No. 5, 1823-1830.

Duncan, R. A., 1969. "Neutral Wind and Universal-Time Control of the Polar F-Region." Journal of Atmospheric and Terrestrial Physics, Vol. 31, 1003-1009.

Elford, W. G., 1959. "A Study of Winds between 80 and 100 km." Planetary and Space Science, Vol. 1, 94-101.

Fogle, B., 1968. "The Climatology of Noctilucent Clouds According to Observations made from North America during 1964-1966." The Meteorological Magazine, Vol. 97, 193-204.

Greenhow, J. S., and E. L. Neufeld, 1959. "Measurements of Turbulence in the 80 to 100 km Region from Radio Echo

Observations of Meteors." Journal of Geophysical Research, Vol. 64, 2129.

Greenhow, J. S., and E. L. Neufeld, 1961. "Wings in the Upper Atmosphere." Quarterly Journal of the Royal Meteorological Society, Vol. 87, 472-489.

Hesstvedt, E., 1968. "On the Effect of Vertical Eddy Transport on Atmospheric Composition in the Mesosphere and Lower Thermosphere." Geofysiske Publikasjoner, Vol. XXVII, No. 4.

Hines, C. O., 1968. "An Effect of Molecular Dissipation in Upper Atmospheric Gravity Waves." Journal of Atmospheric and Terrestrial Physics, Vol. 30, 845-849.

Hines, C. O., 1960. "Internal Atmospheric Gravity Waves at Ionospheric Heights." Canadian Journal of Physics, Vol. 38, 1441-1481.

Hinteregger, H. E., 1961. "Preliminary Data on Solar Extreme Ultraviolet Radiation in the Upper Atmosphere." Journal of Geophysical Research, Vol. 66, No. 8, 2367-2380.

Hodges, R. R., Jr., 1967. "Generation of Turbulence in the Upper Atmosphere by Internal Gravity Waves." Journal of Geophysical Research, Vol. 72, No. 13, 3455-3458.

Hook, J. L., 1970. "Winds at the 75-110 km Level at College, Alaska." Planetary and Space Science, Vol. 18, 1623-1638.

Hunt, D. C., and T. E. Van Zandt, 1961. "Photoionization Heating in the F REgion of the Atmosphere." Journal of Geophysical Research, Vol. 66, No. 6, 1673-1682.

Israel, H., 1957. Atmosphärische Elektrizitat, Volumes I & II, Akademische Verlagsgesellschaft, Geest and Portig K. -G., Leipzig.

Johnson, F. S., 1956. "Temperature Distribution of the Ionosphere under Control of Thermal Conductivity." Journal of Geophysical Research, Vol. 61, 71-76.

Johnson, F. S., 1958. "Temperatures in the High Atmosphere." Annals of Geophysics, Vol. 14, 94-108.

Johnson, F. S., 1965. "Structure of the Upper Atmosphere." In Satellite Environment Handbook, pp 3-22, edited by F. S. Johnson, Stanford University Press, 193 p.

Johnson, F. S., and E. M. Wilkins, 1965. "Thermal Upper Limit on Eddy Diffusion in the Mesosphere and Lower Thermosphere." Journal of Geophysical Research, Vol. 70, 1281-1284 and 4063.

Jones, W. L., 1969. "Atmospheric Internal Gravity Waves and Tides." In Stratospheric Circulation, pp. 469-482, edited by W. L. Webb, Academic Press, Inc., New York, 600 p.

Kallmann Bijl, H. K., 1961. "Daytime and Nighttime Atmospheric Propeties Derived from Rocket and Satellite Observations." Journal of Geophysical Research, Vol. 66, 787-795.

Kashcheyev, B. L., and I. A. Lysenko, 1968. "Atmospheric Circulation in the Meteor Zone." Journal of Atmospheric and Terrestrial Physics, Vol. 30, 903-905.

King, J. W., H. Kohl, D. M. Preece and C. Seabrook, 1968. "An Explanation of Phenomena Occurring in the High-latitude Ionosphere at Certain Universal Times." Journal of Atmospheric and Terrestrial Physics, Vol. 30, 11-23.

Kochanski, A., 1963. "Circulation and Temperature at 70- to 100-Kilometer Height." Journal of Geophysical Research, Vol. 68, No. 1, 213-226.

Kochanski, A., 1964. "Atmospheric Motions from Sodium Cloud Drifts." Journal of Geophysical Research, Vol. 69, No. 17, 3651-3662.

Lauter, E. A., K. Sprenger and G. Entzian, 1969. "The Lower Ionosphere in Winter." In Stratospheric Circulation, pp. 401-438, edited by W. L. Webb, Academic Press, Inc., New York, 600 p.

Leovy, C., 1964. "Radiative Equilibrium of the Mesosphere." Journal of Atmospheric Sciences, Vol. 21, No. 3, 238-248.

Lettau, H., 1951. "Diffusion in the Upper Atmosphere." In Compendium of Meteorology, pp. 320-333, edited by T. F. Malone, American Meteorological Society, Boston, 1334 p.

Lindzen, R. L., 1967. "Thermally Driven Diurnal Tide in the Atmosphere." Quarterly Journal of the Royal Meteorological Society, Vol. 93, 18-42.

Lysenko, I. A., 1963. "Air Currents in the Meteor Zone According to Radio Echo Observations." Soviet Astronomy, Vol. 7, 121-128.

Maeda, K., 1964. "On the Acoustic Heating of the Polar Night Mesosphere." Journal of Geophysical Research, Vol. 69, No. 7, 1381-1395.

Maeda, K., and H. Maeda, 1969. "Stratosphere-Ionosphere Coupling." In Stratospheric Circulation, pp. 339-390, edited by W. L. Webb, Academic Press, Inc., New York, 600 p.

Matsushita, S., 1965. "Global Presentation of the External Sq and L Current Systems." Journal of Geophysical Research, Vol. 70, No. 17, 4395-4398.

Mauchly, S. J., 1923. "Diurnal Variation of the Potential Gradient of Atmospheric Electricity." Journal of Terrestrial Magnetism and Atmospheric Electricty, Vol. 28, 61-81.

Miers, B. T., 1965. "Wind Oscillations Between 30 and 60 km over White Sands Missile Range, New Mexico." Journal of Atmospheric Sciences, Vol. 22, No. 4, 382-387.

Morris, J. E., and M. D. Kays, 1969. "Circulation in the Arctic Mesosphere in Summer." Journal of Geophysical Research, Vol. 74, No. 2, 427-434.

Murphy, C. H., G. V. Bull and H. D. Edwards, 1966. "Ionospheric Winds Measured by Gun-Launched Projectiles." Journal of Geophysical Research, Vol. 71, No. 9, 4535-4544.

Newell, R. E., J. R. Mahoney and R. W. Lenhard, 1966. "A Pilot Study of Small Scale Wind Variations in the Stratosphere and Mesosphere." Quarterly Journal of the Royal Meteorological Society, Vol. 92, 41-54.

Nicolet, M., 1960. "Les Variations de la Densité de la Densité et du Transport de Chaleur par Conduction dans l'atmosphère Supérieure." In Space Research, pp. 46-89, edited by H. Kallmann Bijl, North-Holland Publishing Company, Amsterdam.

Pitteway, M. L. V., and C. O. Hines, 1963. "The Viscous Damping of Atmospheric Gravity Waves." Canadian Journal of Physics, Vol. 41, 1935-1948.

Reed, R. J., 1962. "Some Features of the Annual Temperature Regime in the Tropical Stratosphere." Monthly Weather Review, Vol. 90, 211-215.

Rosenberg, N. W., and C. G. Justus, 1966. "Space and Time Correlations of Ionospheric Winds." Radio Science, Vol. 1, No. 2, 149-155.

Spitzer, L., Jr., 1952. "The Terrestrial Atmosphere above 300 km." In Atmospheres of the Earth and Planets, pp. 211-247-, edited by G. P. Kuiper, University of Chicago Press.

Wagner, C. -U., 1968. "About a "Semi-Annual" Variation of the Amplitude of the Geomagnetic Sq-Variations in Median Latitudes." Journal of Atmospheric and Terrestrial Physics, Vol. 30, 579-589.

Wagner, C. -U., 1969. "The "Semi-Annual" Variation of the Solar Daily Quiet Geomagnetic Variation in the European Region." Gerlands Beiträge zur Geophysik, Vol. 78, No. 2, 120-130.

Webb, W. L., 1964. "Stratospheric Solar Response." Journal of Atmospheric Sciences, Vol. 21, No. 6, 582-591.

Webb, W. L., 1965. "Scale of Stratospheric Detail Structure." In Space Research V, pp 997-1007, edited by D. G. King-Hele, P. Muller and G. Righini, North-Holland Publishing Company, Amsterdam, 1248 p.

Webb, W. L., 1966a. "Stratospheric Tidal Circulations." Reviews of Geophysics, Vol. 4, No. 3, 363-375.

Webb, W. L., 1966b. "The Stratospheric Tidal Jet." Journal of Atmospheric Sciences, Vol. 23, No. 5, 531-534.

Webb, W. L., 1966c. Structure of the Stratosphere and Mesosphere, Academic Press, Inc., New York, 382 p.

Webb, W. L., 1967. "Circulation of the Middle and Upper Atmosphere." In Space Research VII, pp. 1-19, edited by R. L., Smith-Rose, S. A. Bowhill and J. W. King, North-Holland Publishing Company, Amsterdam, 1479 p.

Webb, W. L., 1968. "Source of Atmospheric Electrification." Journal of Geophysical Research, Vol. 73, No. 16,pp. 5061-5071.

Webb, W. L., 1969 (editor). AIAA Progress in Astronautics and Aeronautics: Stratospheric Circulation, Vol. 22, Academic Press, Inc., New York, 600 pp.

Webb, W. L., 1970. "The Cold Earth." In Space Technology and Earth Problems, Volume 23, pp. 85-96, American Astronautical Society, Tarzana, California.

Weise, H., 1951. "Anomalien des Täglishen Ganges im Erdmagnetismus und ihr Zusammenhang mit den Windströmungen der Tiefen Ionosphäre." Zeitschrift für Meteorologie, Vol. 5, No. 12, 373-377.

Whipple, F. J. W., 1929. "On the Association of the Diurnal Variation of Electric Potential Gradient in Fine Weather with the Distribution of Thunderstorms over the Globe." Quarterly Journal of the Royal Meteorological Society, Vol. 55, 1-17.

CHAPTER 2

LOWER IONOSPHERIC STRUCTURE

James E. Midgley

University of Texas at Dallas

2.1 Electron Production

A quantitative description of the ionosphere was first developed in 1931 by Chapman (1931). If the atmospheric molecules have an absorption cross section $\sigma(h)$ for a mono-chromatic beam of ionizing radiation, then the *optical depth* of the atmosphere at a particular altitude h (Figure 2.1) for radiation has a zenith angle χ at that altitude which is

$$\tau(h,\chi) = \int_h^\infty \sigma(h') \, n(h') \, \sec\chi' \, dh' \qquad (2.1)$$

Note that $h'\sin\chi' = h \sin\chi$, if h is measured from the center of the earth.

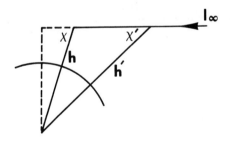

Fig 2.1 Absorption geometry.

The intensity of the radiation at that altitude is then

$$I = I_\infty e^{-\tau}$$

If η ions are formed for each unit of radiation absorbed, the rate of ion production is

$$q(h,\chi) = \eta\sigma n I_\infty e^{-\tau} \qquad (2.2)$$

This peaks at that level where the relative increase of σn is just canceled by the relative decrease of $e^{-\tau}$.

$$\frac{1}{\sigma n}\frac{\partial}{\partial h}(\sigma n) = \frac{\partial\tau}{\partial h} \qquad (2.3)$$

If χ is *small enough* that $\sec\chi$ does not change appreciably through the production region, it can be factored out of the optical depth integral and the height of maximum production, h_{max} (Fig. 2.2) can be written more simply as

$$\frac{\partial}{\partial h}\left(\frac{1}{\sigma n}\right) = \sec\chi \qquad (2.4)$$

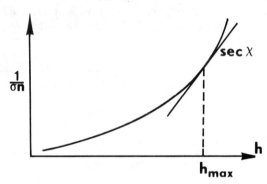

Fig. 2.2 Determination of the height of maximum production from a plot of the transparency of the atmosphere.

If, in addition, σ/mg = constant and we introduce the reduced height

$$z = \int_{h_o}^{h}\frac{dh'}{H} \qquad (2.5)$$

where $H = kT/mg$ and h_o is chosen as the altitude where $n_o\sigma_o H_o = 1$, then the optical depth integral can be done analytically. Using the hydrostatic equation, $dp = -nmg\,dh$, the ideal gas law, $p = nkT$, and the constancy of $\sigma H/T$

$$\sigma n H = e^{-z} \qquad (2.6)$$

so the optical depth may be written

$$\tau(z,\chi) = \sec\chi e^{-z} \qquad (2.7)$$

and the production function is

$$q(z,\chi) = \frac{nI_\infty}{eH(z)} \exp(-z+1-e^{-z}\sec\chi) \qquad (2.8)$$

This function has a maximum at $z_m = \ln\left[\dfrac{\sec\chi}{1+dH/dh}\right]$ where its value is

$$\frac{nI_\infty}{eH(z)} \cos\chi(1+dH/dh)e^{-dH/dh} \qquad (2.9)$$

If H were constant, the *normalized* production function would appear as shown in Figures 2.3 and 2.4, plotted on linear and logarithmic scales, respectively.

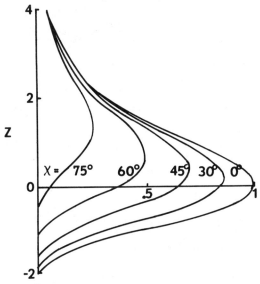

Fig. 2.3 Chapman Production Function for a uniform isothermal atmosphere at various solar zenith angles (Rishbeth & Garriott, 1969).

If $\chi \approx 90°$, the *variation* of $\sec\chi$ through the production layer cannot be ignored. In such a case the optical depth integral is better written with χ' as the variable

$$h' = \frac{h \sin\chi}{\sin\chi'} \qquad \sec\chi'dh' = \frac{-h \sin\chi d\chi'}{\sin^2\chi'}$$

$$\tau(h,\chi) = \int_0^\chi \sigma n\left[\frac{h\sin\chi}{\sin\chi'}\right] h \sin\chi \frac{d\chi'}{\sin^2\chi'} \qquad (2.10)$$

If, in addition, σ and H are constant through the pro-

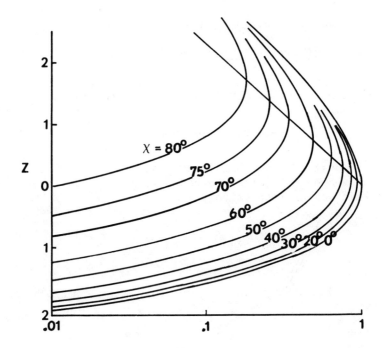

Fig. 2.4 The Chapman Production Function plotted on a logarithmic scale.

duction layer, this integral may be simplified

$$\tau(h,\chi) = \sigma Hn(h)Ch(h/H,\chi) \qquad (2.11)$$

where $Ch(R,\chi) = \int_0^\chi \exp(R\{1-\frac{\sin\chi}{\sin\chi'}\})\ R\ \sin\chi\ \frac{d\chi'}{\sin^2\chi'}$ (2.12)

is known as the *Chapman function*. It simply replaces $\sec\chi$
in(2.8)when H and σ are constant. The Chapman function has
been tabulated by Wilkes (1954) in considerable detail
and it is plotted in Fig. 2.5 for several h/H. For χ near
$90°$ it may be approximated by

$$Ch(R,\chi) = \sqrt{\frac{\pi}{2}R\sin\chi}\ e^x\{1+\sqrt{erf\ (x)}\}$$

where $x = \frac{1}{2}R\cos^2\chi$

In the actual atmosphere the radiation is not *monochromatic* and it is necessary to integrate the production function
over the spectrum. Both σ and η depend on wavelength λ as
well as the type of gas being ionized so the actual production

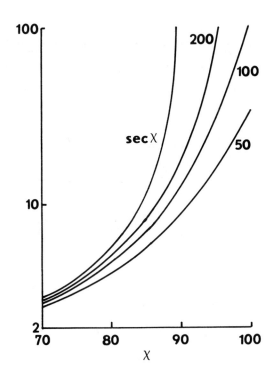

Fig. 2.5 Chapman function versus zenith angle for various
values of the ratio of geocentric distance to scale height.

function must be written

$$q(h,\chi) = \int d\lambda\, I_\infty(\lambda) \sum_i \eta_i(\lambda)\sigma_i(\lambda)n_i(h) \; \exp\left[-\sum_j \int_h^\infty \sigma_j(\lambda)n_j(h')\sec\chi'dh'\right]$$

$$(2.13)$$

where the summations are over the various absorbing gases.
Exact evaluation of this would be a very involved numerical
calculation, but it is generally simplified first by means of
the sort of approximations already discussed.

    To a fair degree of approximation it is found that for
short wavelength one ion is formed for each 34 ev of ionizing
radiation. Thus Eq. 2.13 may be appropriately simplified
by replacing the factor $I_\infty(\lambda)\eta d\lambda$, representing the total
number of ions/cm^2/sec produced by the radiation between λ and
$\lambda+d\lambda$, by the factor $F_\infty(\lambda)d\lambda$, representing the energy flux
(in units of 34ev) in the radiation between λ and $\lambda+d\lambda$.

$F_\infty(\lambda)$ is a rapidly varying function containing many bands
and discrete lines. A smoothed version drawn from data given
by Allen (1965) is illustrated in Fig. 2.6. The strongest
lines are not smoothed into the continuum but drawn with a
uniform 10 Angstrom width so that their relative strengths
are apparent.

The *absorption cross section* for N_2 and the altitude of
unit optical depth $h_o(\lambda)$, where

$$\sum_i \sigma_i(\lambda) n_i(h_o) H_i(h_o) = 1 \qquad (2.14)$$

is also shown in Fig. 2.6. The absorption cross section for
O_2 is quite similar to that for N_2. Beyond 700 Å both vary
so erratically that the height of unit optical depth becomes
very uncertain, and the dashed part of the curve represents
only its approximate average. For instance, for $N_2 \sigma(H_\gamma=973A)$
/ $\sigma(C\ III=977Å) = 100$, so even though H_γ reaches unit optical
depth at 230km, CIII does not do so until 110km.

In general, the most strongly absorbed radiation, 170-800
Å, deposits most of its energy between 140 and 170 km, forming
the F1 layer as is indicated in Fig. 2.7. Radiation below
100 Å and in the range 800-1027 Å provide the E region
ionization between 100 and 120 km. H_α, at 1216 Å , is the
strongest line in the UV spectrum. Lying above the ionization
limit of N_2, O, and O_2, it mostly just dissociates O_2 but at
70-80 km the trace constituent NO becomes sufficiently concentrated to contribute significant ionization, forming the
D region. Below 60 km the atmospheric density has increased
sufficiently that the major source of ions is cosmic rays;
this is called the C layer. When a flare occurs on the sun,
the intensity of high energy X rays may be increased by many
orders of magnitude, and the electron production in the D
region increased by several orders of magnitude.

2.2 Electron Loss

An electron cannot *recombine* with an ion unless some
other particle is involved to carry away the excess energy.
This could be a photon emitted by the reaction, but the rates
for such radiative recombinations are very small, typical
lifetimes being over ten days even for N_e as high as 10^6.
It could be another molecule, but such high particle concentrations are required for these three body reactions that
they are as slow as radiative recombination at 70 km, and
they decrease with height as the neutral density. The

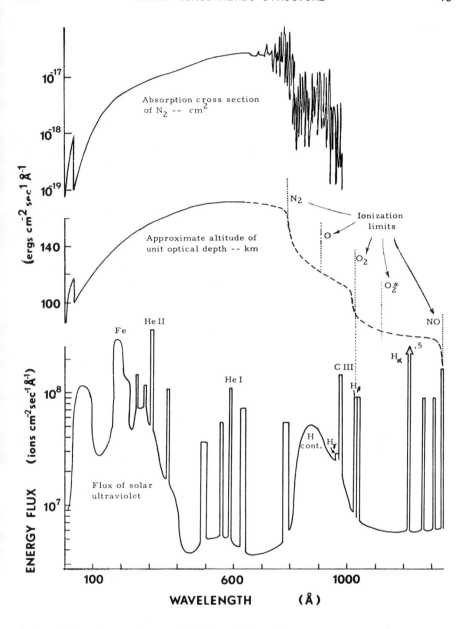

Fig. 2.6 Energy Flux of Solar Ultraviolet as a function
of wavelength (Allen, 1965). The strongest lines are not
smoothed into the continuum, but drawn with a uniform 10
Angstrom width so that their relative strengths are apparent.
The absorption cross section of N_2, the principle absorber,
is also shown along with the approximate altitude at which
each wavelength reaches unit optical depth.

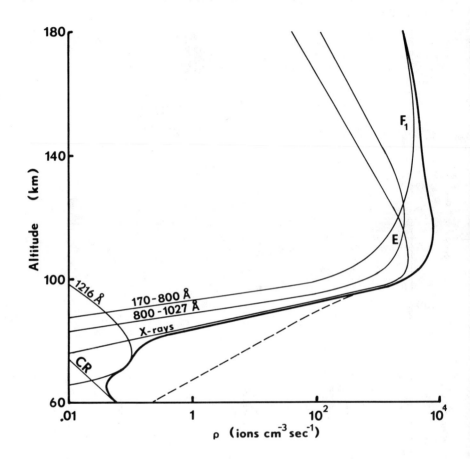

Fig. 2.7 Shape of the ion production layers corresponding to various wavelength bands.

required particle could be a part of the ion itself, if it is a molecular ion, and since the rates for such dissociative re-combinations are about 10^5 greater than for radiative recom-bination, this is the primary mechanism of recombination in the ionosphere.

There is one other loss process which is as rapid as dissociative recombination: *ion-ion recombination*. The im-portance of this process depends on the ratio of negative ions to electrons (Fig. 2.8). These ions are formed by radiative attachment (or collisional attachment below 70 km) to neutrals (NO_2, O_3, O_2 and O). During the day they have a lifetime of only about one second (Fig. 2.9) before being destroyed by photodetachment. At night the primary loss is collisional detachment by molecules in metastable states.

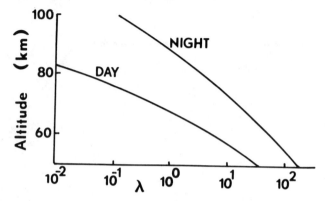

Fig. 2.8   Approximate values of the negative ion ratio in the normal D region {calculated by Rishbeth and Garriott (1969) from data given by Nicolet and Aikin}.

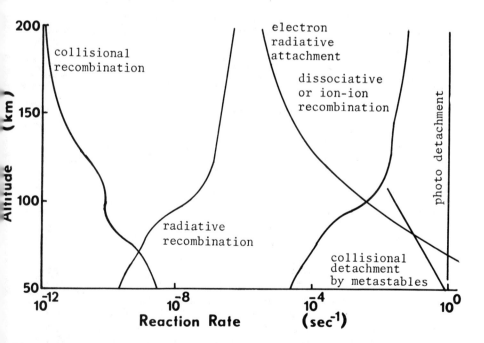

Fig. 2.9   Typical reaction rates for various types of ionic reactions as a function of altitude.

Above 90km essentially all electron loss is by the dissociative recombination of NO^+ and O_2^+, and until H^+ becomes significant at about 500 km the chemistry of the atmosphere and ionosphere involves N, O and their compounds almost exclusively. All of the reactions that are ever of any importance are summarized in the following chart (Fig. 2.10).

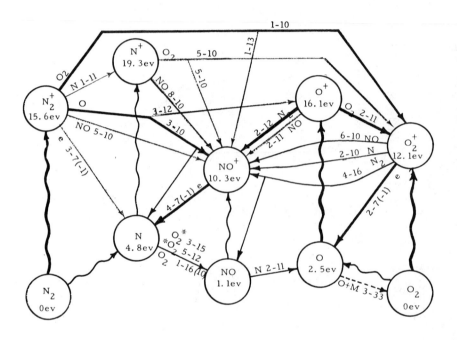

Fig. 2.10 Chemistry of Nitrogen and Oxygen Ions. Wavy lines represent photoionization or photodissociation. Straight lines are reactions, and the numbers thereon give the rate as explained in the test.

The species are plotted vertically in accordance with their relative energy. The rate constants are shown for each reaction in abbreviated form: 4-7(-1) means $=4 \cdot 10^{-7} (T/300)^{-1}$ cm^3/sec. The temperature dependence of most reactions is not shown because it is either unknown or constant. These rates were obtained mostly from Keneshea et al. (1970) and Ferguson (1967). The most important reactions are shown as heavy lines and the unimportant reactions as very light lines. The most complete summary of reaction rates is in ESSA Technical Report ERL 135-AL 3, published by the US Government Printing Office in Sept 1969.

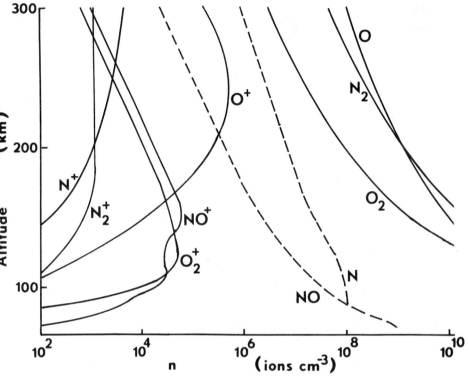

Fig. 2.11 Neutral and ionic concentrations during the daytime
solar minimum accòrding to Johnson (1967). Dashed curves are
estimates.

In order to understand the relative importance of the
possible reactions, it is necessary to have some knowledge of
the neutral concentrations. These are shown approximately in
Fig. 2.11.

In the range 130-260km N_2^+ is the ion produced most frequently. It could recombine dissociatively in about 30 seconds, but most of it reacts with O to form NO^+ in a time
decreasing from 13 seconds at 260km to 0.1 seconds at 130 km.
At 100km it will react with O_2 in 14 milliseconds, and below
that altitude this is the primary loss process. In short, the
primary ions below 500km are O^+, NO^+, and O_2^+.

Under equilibrium conditions in the D and E region the
production and loss at a given point must balance out. Considering only those reactions shown as heavy lines on the
chart, the concentrations of N_2^+ and O^+ may be written immediately since they depend only on the photo-production rates
and the concentrations of N_2, and O, and O_2.

$$\{N_2^+\} = \frac{q(N_2^+)10^{10}}{3\{O\}+\{O_2\}} \qquad \{O^+\}= \frac{q(O^+)5\cdot10^{11}}{\{N_2\}+ 10\{O_2\}}$$

The concentration of NO^+ and O_2^+, however, depend on the total electron concentration since they are lost by electron re-combination:

$$4\cdot10^{-7}\{NO^+\}\{e\} = 3\cdot10^{-10}\{N_2^+\}\{O\} + 2\cdot10^{-12}\{O^+\}\{N_2\}$$

$$2\cdot10^{-7}\{O_2^+\}\{e\} = 10^{-10}\{N_2^+\}\{O_2\}+ 2\cdot10^{-11}\{O^+\}\{O_2\}+ q(O_2^+)$$

Equating the sum of all these ion concentrations to electron concentration gives a single equation for electron concentration:

$$e^2=e(N_2^++O^+)+N_2^+(.75\times10^{-3}O+.5\times10^{-3}O_2)+O^+(.5\times10^{-5}N_2+10^{-4}O_2+5\times10^6qO_2^+)$$

In the lower E region where neutral concentrations are large, $\{N_2^+\}$ and $\{O^+\}$ are small and the first term on the right may be neglected. Then $\{e\} =\{NO^+\} +\{O_2^+\}$ and the loss rate is proportional to the square of $\{e\}$. That part of the F region where this is the case is called the F_1 region. As the neutral densities decrease, however, eventually the first term dominates the second and $\{e\} = \{O^+\} + \{N_2^+\}$. This is the F_2 region.

2.3 Transport of Ionization

Until now the development has assumed that electrons and ions stayed near where they were formed until they recombined. This is a good approximation in the E region for the mean free paths of ions are only about 8 cm at 100 km. The mean free path increases with altitude, however, until it is about 200 m at 200 km altitude, and 2 km at 300km. In addition the lifetimes of the ions are increasing with altitude, being about 4 minutes for O^+ at 200km and 100 minutes at 300km. Thus above 250km the ions can move a long way during their lifetime and they tend to come into diffusive equilibrium. If it were not for this effect the formula at the top of the page shows that $\{O^+\}$ would increase indefinitely since the production is decreasing less rapidly than the loss.

Diffusive equilibrium of a plasma is quite different from diffusive equilibrium of a neutral gas. If the electrons were not charged, their scale height would be about 30,000 times greater than the scale height for O at the same temperature. But even slight differences in the spatial distribution

of the electrons and ions will set up such strong electric
fields that it is an excellent approximation to say that
they must have exactly the same distribution. In a steady
state, then, the electric field must support half the weight
of the average ion and increase the downward force on the
electrons by a like amount, so that both the ions and the
electrons have the same effective mass. Alternatively, one
might think of the electron-ion pair as a single particle with
the ion mass but twice the energy, i. e., twice the temperature.

This turns out to be the most useful way to think of it,
for whether the plasma is in equilibrium or not, each ion must
have on the average exactly one electron in its immediate
vicinity. Consider the steady state equations for the vertical momentum of each species: (note: $f' = df/dz$)

$$(n \, k \, T_e)' \; = \; -n \, e \, E$$

$$(n_i k T_i)' \; = \; n_i \, (eE - m_i g - m_i w_i \nu_i)$$

$$i = 1, \, 2, \, 3, \, \ldots \ldots ,$$

m_i, w_i and ν_i are the mass, mean vertical velocity, and collision frequency with neutrals of the i th species.

The pressure associated with the i th species and its
share of the electrons is $P_i = n_i k (T_i + T_e)$ and the local
scale height of that species is $H_i = -P_i/P_i{}'$. Assume that
T_i is the same for all ions so that we may define an "effective plasma temperature" $T = T_i + T_e$ and total pressure
$P = nkT$.

Summing the equations for all species gives the simple
equation

$$P' \; = \; -mn \, (g+w\nu)$$

where the following average quantities are defined: $m =$
$\Sigma m_i n_i/n$, $w = \Sigma m_i n_i w_i/mn$, $\nu = \Sigma m_i n_i w_i \nu_i/mnw$. This equation may
be rewritten

$$w = \frac{kT}{m\nu} \left[\frac{1}{H} - \frac{1}{\overline{H}} \right]$$

where $\overline{H} = kT/mg$ is the equilibrium value of H, the scale height
of the total plasma. Of course the magnitude of the velocity
depends strongly on ν, and its importance depends on how far
a particular particle moves during its lifetime.

Some interesting effects become apparent if we look at
the diffusion of a single ionic constituent--particularly a
minor constituent.   Sum the ion equations

$$kT_i \; n'/n + kT_i' = eE - mg - mw\nu$$

and eliminate n'/n using the electron equation.

$$kT_e T_i' - kT_i T_e' = eET - mT_e(g+w\nu)$$

This equation may now be used to eliminate eE from the equation
for the i th ion.

$$w_i = w\frac{\nu m T_e}{\nu_i m_i T} + \frac{kT_i}{m_i \nu_i}\left[\frac{1}{H_i} - \frac{1}{\overline{H}_i}\right]$$

where

$$\overline{H}_i = \frac{kT}{m_i g}\left[1 + \frac{T_e}{T_i}(1 - \frac{m}{m_i})\right]^{-1}$$

is the equilibrium scale height of this ion.   Clearly if $m_i = m$
this constituent simply has the local plasma scale height.
If, however, $m_i < mT_e/T$ the scale height will be negative; and,
if $m_i \gg m$, the scale height will be only slightly greater than
the corresponding neutral scale height.   Both these effects
are well illustrated in the concentration of H^+ and O^+ between 400km and 1200km shown in Fig. 2.12 taken from Istomin
(1966).

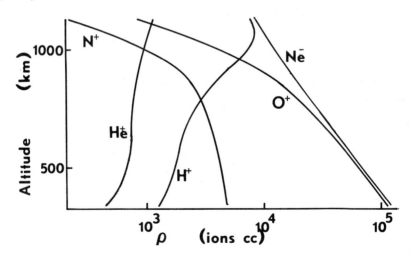

Fig. 2.12 Typical variation of ionic concentrations in the
region where the light ions become dominant.

If the light minor constituent is escaping, as in the polar regions, $n_i w_i$ must be constant, so $n_i \sim v_i$, and it will have approximately the scale height of the neutrals.

2.4 Time Variations

We have seen how production, loss, and transport all affect the density of electrons at any one point. If they do not balance out at any instant, then the electron density will vary with time according to the equation:

$$\frac{\partial n}{\partial t} = P - L - \nabla \cdot (n\mathbf{w})$$

Naturally the major variation is the diurnal one, since the production function essentially goes to zero at night. Profiles of O^+- O_2^+, and NO^+ during the day, at sunset and at night (taken from a report given by C Y Johnson at a COSPAR Symposium in 1967) are shown in Fig. 2.13.

Many of the features of the decay at sunset are explainable by the fact that the reaction of N_2^+ with O is the major source of NO^+, while the reaction of N_2^+ with O_2 is only a secondary source of O_2^+. Thus with the disappearance of N_2^+ at sunset

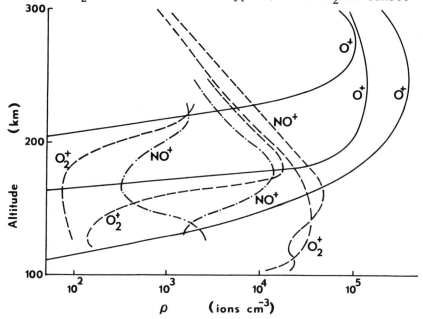

Fig. 2.13 Profiles of O^+, O_2^+, and NO^+ during the day, at sunset, and at night.

NO^+ reduces by a factor of 3 immediately even in those regions where there is sufficient O^+ to maintain the O_2^+ at essentially its daytime concentration. As the O^+ is gradually removed by the formation of the molecular ions the crossover between NO^+ and O_2^+ moves to greater altitudes. At the lower altitudes the NO^+ dominates by an order of magnitude, because the lifetime of O_2^+ for conversion to NO^+ is shorter than the lifetime of NO^+ for dissociative recombination at the low nighttime electron densities.

For the lifetime of an NO^+ to exceed eight hours (as it must at night in the E region if there were no source of additional ionization) the electron density would have to be less than 100/cc. Clearly there must be some nighttime ionization around 100km of about 2 electrons/cc/sec, and this is assumed to be scattered Lyman β.

2.5 References

Allen, C. W., 1965. "The Interpretation of the XUV Solar Spectrum." Space Science Reviews, Vol. 4, 91.

Chapman, S., 1931. "The Absorption and Dissociative or Ionizing Effect of Monochromatic Radiation in an Atmosphere on a Rotating Earth." Proceedings of the Physical Society, Vol. 43, 26, 483.

Ferguson, E. E., 1967. "Ionospheric Ion-Molecule Reaction Rates." Reviews of Geophysics, Vol. 5, 3, 305.

Istomin, V. G., 1966. "Observational Results on Atmospheric Ions in the Region of the Outer Ionosphere. Annales de GeoPhysique, Vol. 22, 255.

Johnson, C. Y., 1967. "Ion and Neutral Composition of the Ionosphere." IQSY/COSPAR Joint Symposium, Imperial College, London.

Keneshea, T. J., R. S. Narcisi and W. Snider, Jr., 1970. "Diurnal Model of the E Region." Journal of Geophysical Research, Vol. 75, 845.

Rishbeth, H., and O. Garriott, 1969. Introduction to Ionospheric Physics, Academic Press, Inc., New York.

Wilkes, M. V., 1954. "A Table of Chapman's Grazing Incidence Integral." Proceedings of the Physical Society of London (Series B), Vol. 67, 304.

CHAPTER 3

THE TOPSIDE IONOSPHERE

Walter J. Heikkila

The University of Texas at Dallas

3.1 Introduction

The balance between ionization production and loss processes leads to a maximum electron density in the F region
of the earth's ionosphere at an altitude of about 300 km.
Above this maximum the electron density decreases monotoni-
cally up to very great altitudes. The dominant positive ion
is 0+ up to about 1000 km, and then H+ out to the outer limits
of the ionosphere, although He+ can at times be a significant
constituent in the 1000 to 3000 km range.

Two features of this topside ionosphere are of overwhelm-
ing importance in determining its general characteristics.
One is the *long mean free path* of the particles; this implies
large transport coefficients, particularly thermal and electrical conductivity. The other feature is the earth's magnetic field, which introduces strong *anisotropies* in the local
plasma properties and also certain characteristic latitudinal
variations in the morphology. The geomagnetic field also plays
a crucial role in determining the outer boundary of the earth's
ionosphere; this outer ionospheric region is called the magnetosphere.

3.2 The Geomagnetic Field

The *direction seeking* properties of small natural magnets
such as lodestone were known thousands of years ago. However,
it was only 400 years ago that the earth was recognized as the
responsible agent, in the form of a giant magnet. William
Gilbert pointed out that the earth's magnetic properties were
similar to those of a uniformly magnetized sphere, which is a
magnetic dipole.

The geomagnetic field as observed at any point on earth
can be expressed as a vector. However, it is usual to express
the field in terms of components, and there are several schemes

53

in use by which this is done. The various parameters that may
be used are designated magnetic elements, and some of the more
common of these are listed below:

F - Total magnetic intensity.
D - Declination or magnetic variation. This is the angle
    in the horizontal plane between true north and mag-
    netic north (i.e., it is the direction of the horizon-
    tal component of the magnetic field). The declination
    is positive when magnetic north is east of true north.
I - Inclination or magnetic dip. This is the angle be-
    tween the direction of the magnetic field and the
    horizontal plane, taken positive when the field direc-
    tion is downward.
H - Horizontal component of the geomagnetic field inten-
    sity, always regarded as positive.
V - Vertical component of the geomagnetic field intensity.
    Its sign is the same as that of I, and hence it is
    positive when directed downward.
X,Y,Z - The north, east, and vertical components of the
    geomagnetic field intensity. Z is the same as V, and
    hence its positive direction is also downward.

The direction of the magnetic field is that indicated by the
north pole of a compass. The geomagnetic field at any point
can be described by several combinations of the magnetic elements, for example, (X,Y,Z), (F,D,I), (H,V,D), etc.

In geomagnetic work, one does not usually observe the dis-
tinction between magnetic field and magnetic induction. The
field intensity is usually specified in gauss (a cgs unit) or
in gammas; the gamma is 10^{-5} gauss. The relative permeability
of the atmosphere is very nearly unity, so the failure to dis-
tinguish between field strength and induction is tolerable.
In space problems involving magnetic fields, it is common to
use H or B instead of F to describe the field.

The total surface field is shown in Fig. 3.1 and the inclination in Fig. 3.2 (taken from the Satellite Environment
Handbook, F. S. Johnson, ed). Gross features agreeing with
the simple *dipole model* are:

(1) The 2 to 1 increase in intensity from the equator to
    the pole,
(2) horizontal direction (I = $0°$) at the equator,
(3) vertical direction at the poles (I = $90°$ at north
    pole and I = $-90°$ at the south pole), and
(4) northward direction (not shown).

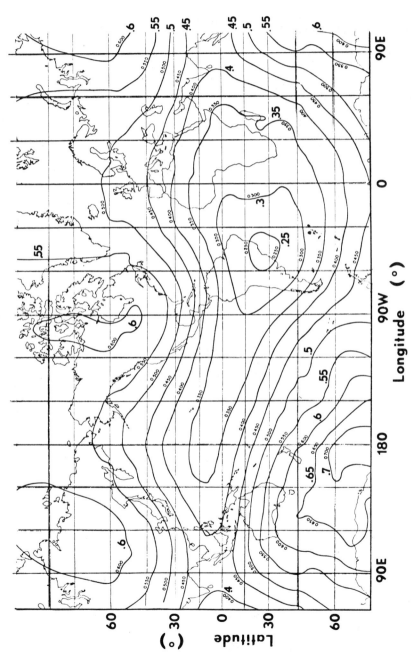

Fig. 3.1 Total intensity F of the Earth's magnetic field (in gauss). *Air Force Geophysics Research Directorate (1960), reproduced with permission of Macmillan Co.*

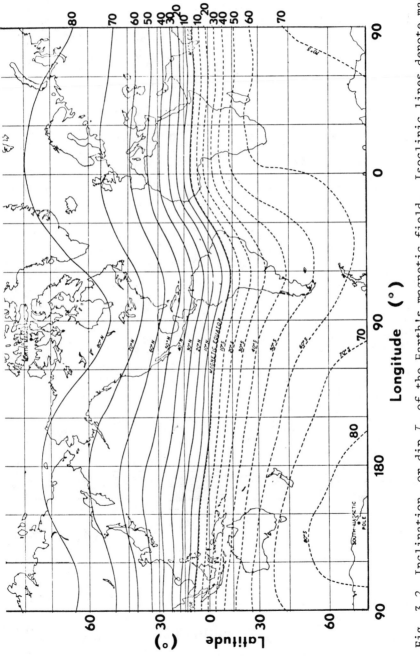

Fig. 3.2 Inclination, or dip I, of the Earth's magnetic field. Isoclinic lines denote magnetic inclination or dip in degrees.

The field components, according to the dipole model, are given in terms of the magnetic moment $M = 8.07 \times 10^{25}$ gauss cm^3 (or maxwell cm), the radial distance R, and the geomagnetic latitude λ, by the relations

$$H = \frac{M \cos \lambda}{R^3} \qquad (3.1)$$

$$Z = \frac{2M \sin \lambda}{R^3} \qquad (3.2)$$

$$\tan I = Z/H = 2 \tan \lambda.$$

For a best fit, the centered-dipole points 11° away from the south pole, producing geomagnetic poles at 79°N, 69°W and 79°S, 111 E.

Noticeable departures from the dipole model include the so-called *South Atlantic anomaly*, with a rather weak surface field, and irregularities in the lines of constant dip, particularly near the Atlantic Ocean. Many theoretical representations of the geomagnetic field are available. These need continual updating because of *secular variations*.

3.3 Motion of Charged Particles in Electromagnetic Fields

The fundamental equations which apply to electromagnetic phenomena are *Maxwell's equations*, which in mks units have the form

$$\nabla \times \mathbf{E} = -\dot{\mathbf{B}} \qquad (3.3)$$

$$\nabla \times \mathbf{H} = \dot{\mathbf{D}} + \mathbf{J} \qquad (3.4)$$

$$\nabla \cdot \mathbf{D} = \rho_c \qquad (3.5)$$

$$\nabla \cdot \mathbf{B} = 0 \qquad (3.6)$$

where $\mathbf{E}$ is the electric field intensity in volts/m, $\mathbf{H}$ is the magnetic field intensity in amp-turns /m, $\mathbf{D}$ is the electric displacement in coul/m^2, $\mathbf{B}$ is the magnetic induction in webers/ m^2, $\mathbf{J}$ is the current density, amp/m^2, and ρ_c is the charge density in coul/m^3.

Other pertinent equations are the constitutive relations

$$\mathbf{D} = \varepsilon \mathbf{E} \qquad (3.7)$$

$$\mathbf{B} = \mu \mathbf{H} \qquad (3.8)$$

with the material constants (or tensors), ε (dielectric constant, farads /m), and μ (magnetic permeability, henrys /m) being determined by the physical nature of the medium.

The conservation of charge

$$\nabla \cdot \mathbf{J} = - \dot{\rho}_c \qquad (3.9)$$

is implied by the second curl equation (3.4), where $\dot{\mathbf{D}}$ is interpreted as a displacement current.

Finally, we need the *Lorentz force* equation

$$\mathbf{F} = q \, (\mathbf{E} + \mathbf{v} \times \mathbf{B}) \qquad (3.10)$$

to relate the fields to their resultant forces on the charged particles. Here q is the charge in coulombs, $\mathbf{v}$ is the particle velocity m/sec, and $\mathbf{F}$ is the force in newtons.

The electric field produces a constant acceleration

$$\frac{d\mathbf{v}}{dt} = \frac{q\mathbf{E}}{m} \qquad (3.11)$$

The effect of the magnetic field is more complicated, as the acceleration is perpendicular to both the velocity and the magnetic field. The particle trajectory is therefore a circle, and the particle kinetic energy is constant. From the study of circular motion we can equate the centrifugal force to the Lorentz force,

$$\frac{m \, v_\perp^2}{r_c} = q \, v \, |\mathbf{B}| \qquad (3.12)$$

where v is the particle speed perpendicular to $\mathbf{B}$, and r_c is the cyclotron radius, the radius of the particle orbit in the plane perpendicular to $\mathbf{B}$. The velocity $v_{\parallel}$ parallel to $\mathbf{B}$ is unaffected, and the total trajectory is thus a helix. An important quantity is the *cyclotron frequency*

$$\omega_c = \frac{v_\perp}{r_c} = \frac{qB}{m} \qquad (3.13)$$

One further important factor is that of collisions suffered by the particle in its motion. This can be represented crudely as a frictional force $-m\nu\mathbf{v}$, where ν is the *collision frequency*. To see its effect we consider

$$m \frac{d\mathbf{v}}{dt} = - m\nu\mathbf{v} \qquad (3.14)$$

which has the solution

$$\mathbf{v}(t) = \mathbf{v}(o)\ e^{-\nu t} \tag{3.15}$$

Hence, the effect of the collision term is to introduce damping of the particle motion produced by other forces.

Combining all these forces we have the equation of motion for a single charged particle in externally imposed electric and magnetic fields, and including the average frictional effect of collisions

$$m\ \frac{d\mathbf{v}}{dt} = q\ (\mathbf{E} + \mathbf{v} \times \mathbf{B}) - m\nu\mathbf{v} \tag{3.16}$$

The resulting idealized particle trajectories are sketched in Fig. 3.3.

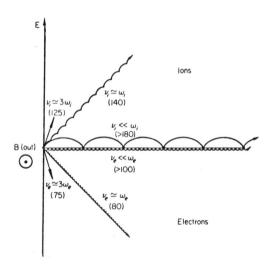

Fig. 3.3　Idealized trajectories for electrons and ions subject to an electric field in the plane of the diagram and a magnetic field directly out of that plane.

In this case charged particles are assumed to collide with neutral particles at regular intervals $1/\nu$ and to possess zero velocity after each collision. In order to show both electronic and ionic gyrations, the diagrams are drawn as though $\omega_e/\omega_i = 10$ (instead of order 10^4 as in reality). All the trajectories refer to equal intervals of time, namely 5

ionic (or 50 electronic) gyroperiods. Numbers in brackets refer to approximate heights, in kilometers, at which the conditions occur. In general, the head of the vector representing V_i or V_e lies on a semicircle.

3.4 Electromagnetic Waves in a Plasma

The above discussion was for a single (or average) particle. An important characteristic of a charged particle gas,
or plasma, is that of *collective action*. This leads to plasma
oscillations whose nature can be seen from a simple argument
originally given by Tonks and Langmuir (1929). We consider
an initially uniform electron gas of density n. By some external means, let a one-dimensional perturbation occur such
that electrons at position χ are displaced in the χ direction
by a small increment $\xi(\chi)$, as shown in Fig. 3.4. The local

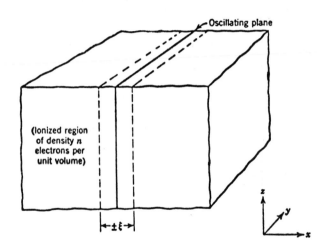

Fig. 3.4 Geometry of one-dimensional perturbation leading to
plasma oscillation.

density of electrons then departs from the uniform density n
by the increment

$$\delta n = n \frac{d\xi}{d\chi} \qquad (3.17)$$

Since the net charge density ρ was originally zero, the perturbed charge density is

$$\delta\rho = -e \ \delta n = ne \ \frac{d\xi}{d\chi} \ , \tag{3.18}$$

where the electron charge is -e. The new net charge is related to the existing electric field by Gauss's law

$$\varepsilon_o \ \frac{dE}{d\chi} = ne \ \frac{d\xi}{d\chi} \tag{3.19}$$

which can be integrated immediately, giving

$$E = \frac{ne}{\varepsilon_o} \ \xi \tag{3.20}$$

within an arbitrary constant. The electric force on each
electron is then

$$F = -eE = - \ \frac{ne^2}{\varepsilon_o} \ \xi . \tag{3.21}$$

For simplicity, we neglect the viscous damping forces which
arise from collisions between the electron and heavy particles.
Newton's equation of motion is then

$$m\xi + \frac{ne^2}{\varepsilon_o} \ \ddot{\xi} = F_{ext}, \tag{3.22}$$

where m is the electron mass, $\ddot{\xi} \equiv \partial^2 \xi / \partial t^2$ is the electron acceleration, and F_{ext} is the external force required to produce
the perturbation. If this external force is suddenly removed,
this equation shows that the electrons oscillate about their
equilibrium positions with simple harmonic motion at the
plasma frequency

$$\omega_p = \left[\frac{ne^2}{\varepsilon_o m} \right]^{\frac{1}{2}} \tag{3.23}$$

The interaction of an electron gas with an electromagnetic
wave can be illustrated by the simple case of no magnetic
field and no collisions. The electron thermal velocity is
assumed zero (cold plasma), and the positive ions are assumed
immobile constituting simply a charge neutralizing background.
Then

$$m \ \frac{\partial \mathbf{v}}{\partial t} = - \ e\mathbf{E} \tag{3.24}$$

With the assumption of a harmonic time variation $e^{i\omega t}$ this gives

$$i\omega m\mathbf{v} = -e\mathbf{E} \tag{3.25}$$

The particle current density is

$$\mathbf{j} = -Ne\mathbf{v} \tag{3.26}$$

where for simplicity (linearity) N is the unperturbed electron density. The combination of these two equations yields

$$\mathbf{j} = \sigma\mathbf{E} \tag{3.27}$$

where

$$\sigma = -i\ \frac{Ne^2}{\omega m} \quad . \tag{3.28}$$

In other words, in respect of its interaction with an electromagnetic field, the electron gas behaves as a medium with scalar conductivity. Since this expression is purely imaginary, the medium is in fact more aptly described as a lossless dielectric. Substitution of $\mathbf{j}$ into the right-hand side of the Maxwell equation (3.4)

$$\text{curl } \mathbf{H} = i\omega\varepsilon_0\mathbf{E} + \mathbf{j} \tag{3.29}$$

yields the dielectric constant of the medium in the form

$$\varepsilon = 1 - \frac{Ne^2}{\varepsilon_0 m\omega^2} \tag{3.30}$$

The quantity $Ne^2/(\varepsilon_0 m)$ has the dimensions of the square of an angular frequency; this angular frequency is the *plasma frequency*, ω_p, discussed above. Then this equation is

$$\varepsilon = 1 - \omega_p^2/\omega^2 \quad . \tag{3.31}$$

This simple model of a plasma thus yields a scalar dielectric constant which is less than unity. The propagation of transverse plane electromagnetic waves is possible at high frequencies, $\omega > \omega_p$. Below the plasma frequency the dielectric constant becomes negative and propagation is *not possible*. A radio wave will be reflected when penetrating a plasma with variable density when $\omega = \omega_p$. The expressions require modification when $\mathbf{B}$, $\nu \neq 0$, when heavy ion motions and thermal motions are important.

This reaction provides the basis for radar sounding of

the ionospheric regions. With a swept frequency bottomside sounder reflections will be obtained at varying heights given by the equality of the local plasma frequency and the instantaneous radar frequency. At the peak of the F region the penetration frequency, or critical frequency, f_oF_2 is reached. Similar sounding can be accomplished by *topside sounders*, as has been done with the Alouette and ISIS satellites. Topside soundings have been especially useful for studies of F region morphology.

3.5 Latitudinal and Longitudinal Ionospheric Variations

The large scale ionospheric structure must of course show a dependence on solar zenith angle, with higher electron densities at low latitudes and during the day, and lower densities near the poles and at night. Such gross features do show up, for example in a chart of the peak electron density in the F region ($N_m F_2$), Fig. 3.5.

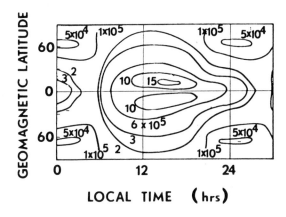

Fig. 3.5 Contours of N_mF_2 in electrons/cm^3.

An interesting feature is apparent near the geomagnetic equator, known as the *Appleton anomaly*. The electron density reaches its greatest values about 10° on either side of the equator, and has a shallow minimum right at the equator (Fig. 3.6). This feature is caused by upward drifting of the ionospheric plasma at the equator due to a westward component of a steady electric field of tidal origin, followed by diffusion down the geomagnetic field lines (Hanson and Moffett, 1966).

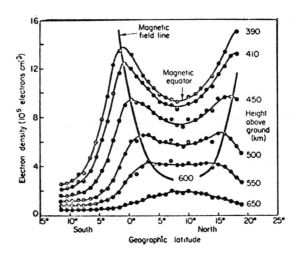

Fig. 3.6 Latitude variation of electron density (electron concentration) at fixed heights, determined by means of the Alouette I topside sounder satellite above Singapore (1° N, 104° E) at 1234 local time, 15 September 1963. A magnetic field line is shown {King et al. (1967)}.

The diurnal variations at the equator are shown in Fig. 3.7 as measured by *incoherent scatter radar* which utilizes the very weak scatter by individual electrons at VHF. On many occasions strong irregularities (indicated by the dots) are present over a wide range of altitudes a few hours after sunset. These are thought to be caused by plasma instabilities. Contours of constant plasma frequency $f_N = \frac{1}{2\pi}\left[\frac{Ne^2}{\varepsilon_o m}\right]^{1/2}$ are

shown in Fig. 3.8 to much higher latitudes. A distinctive feature is a sharp knee in the profiles near 45°N. This feature has been detected by ground based whistler observations and has been called the *plasmapause*, or outer boundary of the plasmasphere (Fig. 3.9). The profiles at 1000 km altitudes (Fig. 3.8) show a plasma trough just beyond this knee. It is thought that electric fields remove the plasma beyond the plasmapause.

The high latitude ionosphere is extremely complex. In addition to photoionization there is a variable and frequently strong *corpuscular ionization,* particularly due to auroral energetic electrons and protons. The maintenance of the

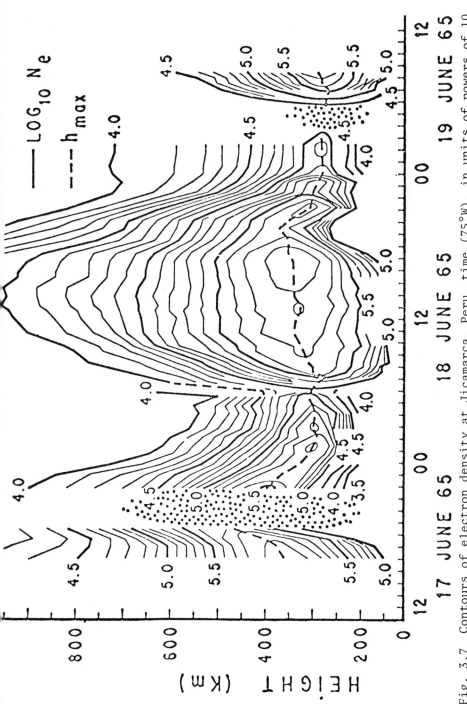

Fig. 3.7 Contours of electron density at Jicamarca, Peru, time (75°W), in units of powers of 10.

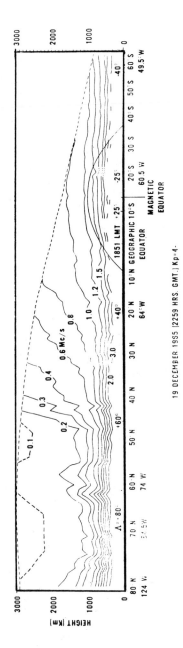

Fig. 3.8 Contours of constant plasma frequency obtained by Nelms and Lockwood (1967) with the Alouette-II topside sounder.

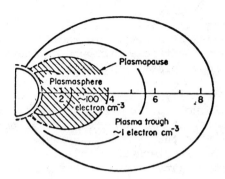

Fig. 3.9 Idealized meridian cross section of the magnetosphere near 1400 hours local time. The shaded region shows the location of the high-density plasma inside the plasmapause or "knee" (electron concentration of order 100 cm^{-3}). The region outside the plasmapause (electron concentration of order 1 cm^{-3}) is linked by magnetic fields lines to the ionospheric "troughs." The dashed part of the boundary shows the low-altitude region in which the structure of the "knee" is not well known. The position of the plasmapause corresponds to moderate but steady magnetic activity (K_p = 2-4) Carpenter (1966) .

F-region ionization during the polar winter is a greater puzzle than even the nighttime F_2 layer. The corpuscular radiation may provide a partial answer, but horizontal and vertical transport may also be important. The escape of plasma along the polar geomagnetic field lines appears to be a reality, the so-called *polar wind*.

3.6   The Magnetosphere

The neutral particles in the earth's upper atmosphere may escape, if they have sufficient energy, in the region above 500 km called the exosphere. Such escape is not possible for the charged particles, because of the presence of the geomagnetic field and the consequent deflection of the moving charged particle by the Lorentz force. The name *magnetosphere* is used for this outer region rather than topside ionosphere to indicate this dominant effect of the geomagnetic field. We need to consider both the thermal plasma in the magnetosphere as well as energetic charged particles.

3.7 Trapped Particles

The main geomagnetic field no longer resembles a dipole field beyond distances of a few earth radii. It is influenced by the flow of charged particles trapped on the field lines and of conducting solar plasma outside.

On the basis of early satellite measurements, Van Allen and Frank (1959) delineated *two radiation belts*, the inner at $\simeq$ 1.5 Re and the outer at $\sim$ 3 to 4 Re. The properties of the belts depend so much on the type and energy of the particles considered that any experimental result is influenced by characteristics of the particle detectors and of the satellite orbit used. This makes it very difficult to summarize the properties of the belts in any simple but meaningful way. It appears, however, that the inner belt contains relatively constant fluxes of *protons* with energies of tens and hundreds of Mev. Beyond 2 Re the flux of protons is smaller, and the outer belt largely consists of energetic *electrons* from a few kev to a few Mev. The fluxes in the outer belt are very variable, and their variations appear to be connected with magnetic disturbances. The ring current which causes the main decrease of magnetic field during storms is believed to be carried by low-energy protons (up to about 50 keV) and by some electrons, trapped at L $\sim$ 4 (Frank, 1967). The trapping region extends to about 6 Re; beyond it, the field configuration does not permit particles to remain in stably trapped orbits, and the auroral particles which populate this region do not belong to the radiation belts proper (Fig. 3.10).

The particles trapped within the radiation belts execute complex trajectories. These can roughly be described as a combination of three types of motion, which are sketched in Fig. 3.11. The numerical values we quote below (Hess et al., 1965) refer to 1 MeV particles at 2000 km altitude at the magnetic equator, though some of the parameters are not very energy dependent.

(1) Gyration around a magnetic field line at the cyclotron period or gyro-period ($2\pi m/eB$), 7µs for electrons and 4 ms for protons, the radius of gyration being 0.3 km and 10 km, respectively.

(2) Bouncing along a field line, between the *mirror points* (one in each hemisphere) at which the particle is reflected in the converging magnetic field. The bounce period is 0.1 s for electrons, 2 s for protons.

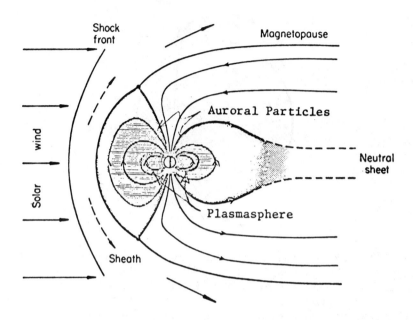

Fig. 3.10 Rough sketch of the earth's magnetosphere, viewed
from the equatorial plane. Solar wind flow is supersonic
where shown by full arrows, but it is subsonic and disturbed
in the apex of the sheath, where shown by dashed arrows. Some
typical geomagnetic field lines are shown, though the extent to
which they merge with the neutral sheet is ill defined. Black
spots indicate two positions where neutral points may exist on
the magnetopause. The domain of auroral particles is stippled,
and the domain of stably trapped particles is shaded.

　　　(3) Drifting azimuthally around the earth, with a period
of revolution of 50 min for electrons, which drift eastward;
30 min for protons, which drift westward.

　　　Each type of motion is characterized by an *adiabatic invariant* quantity, constant for any one particle unless the
particle is subjected to a perturbing force which varies with
a period comparable to the characteristic period of the motion.
In the case of the gyration, the invariant is the magnetic moment of the particle; the other invariants are written in
terms of integrals of the motion. A trapped particle in the
course of bouncing and drifting remains on a magnetic shell
in the region where the field is not grossly distorted from
the dipole form.

　　　Particles are removed from the radiation belts by

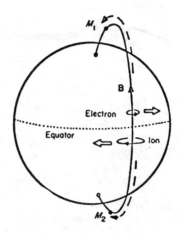

Fig. 3.11 Sketch indicating the motions of energetic particles
trapped on a magnetospheric field line. Particles gyrate about
the lines of magnetic induction **B** in the sense shown by the
small loops. For particles of either kind, the "guiding center" (instantaneous center of gyration) "bounces" back and
forth along the field line between the mirror points M_1, M_2
and "drifts" in longitude as indicated by the hollow arrows.
The longitude drift is partly due to the outward decrease of
$|\mathbf{B}|$ since at any point on the particle trajectory the radius
of gyration is infinitesimally larger on the "outer" part of
the gyration (further from the earth's center) than it is on
the "inner" part; and partly to the curvature of the field
lines.

collisions with neutral particles (Walt and MacDonald, 1964).
Such collisions are extremely rare above 2000 km altitude,
but it is possible for low-frequency electromagnetic waves to
perturb the motions of particles at such high altitudes and
lower their mirror points, thereby causing loss. As a given
particle drifts in longitude, its mirror heights may vary because of variations in the geomagnetic field so that the
rate of loss also varies. It has been suggested that in the
South Atlantic geomagnetic anomaly, where the field is abnormally weak, the loss of radiation belt particles is especially
rapid. This loss might lead to observable phenomena, such as
enhanced airglow and heating and ionization in the ionosphere,
and may even influence the particle distribution in the belts
to a detectable extent.

3.8 The Outer Magnetospheric Region

The *solar wind* (Parker, 1967) consists of charged

particles, especially electrons and protons in the energy
range from 10 ev to 1 kev. It is a plasma, with high electrical conductivity. It interacts strongly with the geomagnetic
field to determine the outer boundary of the magnetosphere,
known as the *magnetopause* (Fig. 3.10). This boundary is
characterized by surface currents caused by the deflection of
charged solar wind particles trying to penetrate into the magnetosphere.

Outside the magnetopause, on the sunward or day side,
exists the shock front or earth's *bow wave* where the supersonic flow of the charged particles is interrupted and slowed
down. Inside the apex of the shock front the flow is disturbed
and subsonic although supersonic flow is resumed further down-
stream.

On the day side, the magnetopause lies at about 10 Re
(earth radii). On the night side, the magnetic field is
drawn out into a long tail (Piddington, 1960) which extends
certainly to the moon's orbit at 60 Re and possibly very much
further. The pattern of field lines within the tail is still
largely unknown. Some field lines may join a weak general
interplanetary field (Dungey, 1961) in which case the tail is
not completely closed. A region of somewhat enhanced plasma
density exists in the tail, and contains a thin *neutral sheet*
in which the magnetic field is very weak, and across which
the field reverses in direction. This sheet is possibly
linked by field lines to the auroral regions. Neutral points
may also exist on the magnetopause, as shown in Fig. 3.10.
The magnetosphere is permeated by hydromagnetic waves and
electromagnetic VLF radiations, which have provided a great
deal of information about its physical condition.

3.9   Ionospheric and Auroral Phenomena in Relation
to the Magnetosphere

Various upper atmospheric phenomena are linked with the
structure of the magnetosphere shown in Fig. 3.10, with the
populations of energetic particles contained in this structure,
and with the circulation of the thermal plasma.

In Fig. 3.10 we showed a domain of *auroral* particles which
lie outside the stable trapping regions populated by the Van
Allen radiation. This domain intersects the earth's surface
in two bands which encircle the two geomagnetic poles. These
bands may be identified with the *auroral ovals* which are the
loci of precipitation of the energetic particles causing
aurora. The ovals are closest to the geomagnetic poles on the

day side and furthest from them on the night side, their approx
imate limits being 78° and 68° dipole latitude under quiet
conditions. In the Northern hemisphere, the ovals correspond
quite well to the area marked with triangles in Fig. 3.12.

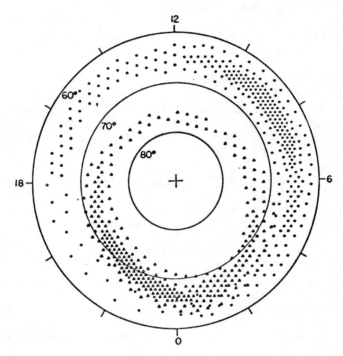

Fig. 3.12 An idealized representation of the two main zones
of auroral particle precipitation in the northern hemisphere,
where the average intensity of the influx is indicated very
approximately by the density of symbols and the coordinates
are geomagnetic latitude and geomagnetic time. The "discrete"
events are represented by triangles (which closely correspond
to the undisturbed "auroral oval") and the "diffuse" events
are indicated by the dots {Hartz and Brice (1967)}. (Geomagnetic time is local solar time measured with respect to geomagnetic longitude.)

The orientation of the auroral ovals is roughly fixed with
respect to the sun. As the earth rotates, each oval appears
to revolve about the geomagnetic poles and, therefore, passes
over different regions of the earth. In particular, let us
consider the midnight sector of the oval, which is the sector
furthest from the pole, and is where bright auroras most often
occur. As the oval revolves, this sector traces out a band,

two or three degrees broad, approximately centered on dipole
latitude 68°. This is the auroral zone. During magnetic disturbance the auroral ovals expand in area, and the auroral
zone moves to a lower latitude (Fel'dshteyn, 1963; Akasofu,
1966).

The distribution of auroras is connected with the distributions of other high-latitude phenomena. The ionospheric
phenomena include radio blackouts (auroral absorption) and
high-latitude sporadic E, whose distributions are obtained
from riometer, radio propagation and ionosonde data. The
occurrence of these phenomena shows marked maxima at certain
times of day, which vary from station to station. Hartz and
Brice (1967) have combined data on many high-latitude phenomena, and plotted them on a polar map in terms of dipole latitude and local geomagnetic time (i.e., longitude with respect
to the direction of the sun); see Fig. 3.12. The phenomena
fall into two classes located in the two different zones postulated by Piddington (1965). One class, "discrete" events
marked by triangles on the map, includes substorm features such
as auroral displays, magnetic bays and ionospheric phenomena
such as auroral absorption, sporadic E and spread F; and
certain types of micropulsation, VLF emission and particle
precipitation. Such phenomena are attributed to intense,
rapidly fluctuating and spatially limited *splashes* of precipitated particles. These events occur in a band which is
very similar (if not identical) to the auroral oval, normally
ranging in latitude from 78° to 68° but expanding towards
lower latitudes during disturbances. The other class, "diffuse"
events marked by dots on the map, includes various steady or
slowly varying auroral, ionospheric and VLF phenomena, which
are attributed to a steady *drizzle* of electrons (>40 kev) from
the outer radiation belt. The drizzle has been measured by
particle detectors on satellites, such as Alouette I (McDiarmid and Burrows, 1964). According to Fig. 3.12, the *diffuse*
events occur in a zone centered at about 65°; they are most
frequent at about 0800 hours local geomagnetic time, whereas
the *discrete* events are most frequent around 2200 hours.
Hartz and Brice state that a similar pattern, but with the
zones shifted poleward by about 5°, applies to the Southern
hemisphere. We must remember that statistical studies of
phenomena do not necessarily display the conditions existing
at any one instant of time.

If the above description is correct, the particles responsible for the phenomena in the auroral oval are not stably
trapped; they may come from the tail region, or even from
outside the magnetosphere via the tail. The low-latitude

boundary of the auroral oval should correspond to the outer
(high-latitude) boundary of the trapped particle domain.

3.10 Soft Particle Measurements

Detailed measurements of *soft particle* precipitation into
the lower atmosphere have been obtained recently by means of
the soft particle spectrometer on the ISIS-I satellite.   The
results for a transpolar pass are shown in Fig. 3.13 in the
form of a spectrogram where the counting rate is used to in-
tensity-modulate a presentation of particle energy vs time.
Occasional sun pulses extending over the entire energy range
10 ev to 10 kev can be identified in the early part of the
record which was obtained at 10:30 local time.   The night side
was traversed at 22:30 local time (Heikkila et al., 1970).

The dayside spin modulated maxima below 100 ev are largely
due to atmospheric photoelectrons which have escaped from the
lower atmosphere along the geomagnetic field lines.   At mid-
latitudes the photoelectron flux shows 2 maxima per spin, corresponding to upgoing and downgoing photoelectrons, as would
be expected on the closed dayside field lines.   Over the pole,
on the other hand, only an upgoing photoelectron flux is seen,
a verification of the open field line structure.

The high latitude limit Λ_c of closed geomagnetic field
lines that are equatorward of the dayside cusp region can be
identified by the sharp cut-off in the high energy (several
kev) electrons at about 75° invariant latitude; this cut-off
is particularly evident in the total energy flux.   The counting rates of more energetic electrons (>20 kev and >40 kev)
also fall to very low values at this latitude.

An intense flux of 100 ev electrons is typically observed
just above this latitude Λ_c, in this case about 10^9 cm^{-2} ster^{-1}
sec^{-1}, extending from 75° to 79°, and carrying up to 0.3 erg
cm^{-2} ster^{-1} sec^{-1}.   The flux appears to be isotropic, with a
total precipitating energy flux equal to about 2 ergs cm^{-2}
sec^{-1}.

A coincident flux of 400 ev protons is also observed at
this location, about two orders of magnitude lower in intensity.

This particle flux appears to be solar wind plasma penetrating via the magnetopause cusp region.   This flux is related to a number of geophysical phenomena, including magnetospheric surface currents, daytime aurora VLF and LF emissions,
ionospheric irregularities, and geomagnetic fluctuations
(Heikkila and Winningham, 1971).

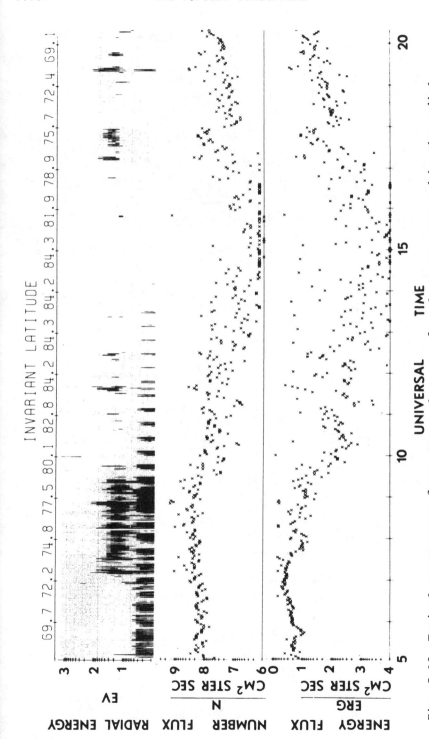

Fig. 3.12 Typical spectrogram for a transpolar pass for electrons measured by the radial detection system.

The nighttime portion of the pass shows fluxes of somewhat more energetic electrons, up to several kev. These
electrons are the primary auroral particles. They appear to
originate in the geomagnetic tail region, where acceleration
processes may be operative. This phenomenon is not well
understood.

It is interesting to note that a nighttime photoelectron
flux was observed on this pass immediately beyond the auroral
zone. This originated from the conjugate southern region,
which was sunlit. This observation confirms that the field
lines just below the nighttime auroral zone are closed.

3.11 References

Akasofu, S. I., 1966. "Electrodynamics of the Magnetosphere:
Geomagnetic Stroms." Space Science Reviews, 6, 21-143.

Carpenter, D. L., 1966. "Whistler Studies of the Plasmapause
in the Magnetosphere. 1. Temporal Variations in the Position
of the Knee and some Evidence on Plasma Motion near the Knee."
Journal of Geophysical Research, 71, 693-709.

Dungey, J. W., 1961. "Interplanetary Magnetic Field and the
Auroral Zones." Physical Review Letters, 6, 47-48.

Farley, D. T., B. B. Balsley, R. F. Woodman and J. P. McClure,
1970. "Implications of VHF Radar Observations." Journal of
Geophysical Research, 75, 7199-7216.

Fel'dshtyn, Ya. I., 1963. "Some Problems Concerning the Morphology of Auroras and Magnetic Disturbances at High Latitudes."
Geomagnetizm i Aeronomiya, English Translation, 3, 183-192.

Frank, L. A., 1967. "Several Observations of Low-Energy Proton
and Electrons in the Earth's Magnetosphere with OGO 3."
Journal of Geophysical Research, 72, 1905-1916.

Hanson, W. B., and R. J. Moffett, 1966. "Ionization Transport
Effects in the Equatorial F Region." Journal of Geophysical
Research, 71, 5559-5572.

Hartz, T. R., and N. M. Brice, 1967. "The General Pattern of
Auroral Particle Precipitation." Planetary and Space Science,
15, 301-329.

Heald, M. A., and C. B. Wharton, 1965. Plasma Diagnostics with
Microwaves, John Wiley & Sons, New York.

Heikkila, W. J., J. B. Smith, J. Tarstrup and J. D. Winningham, 1970. "The Soft Particle Spectrometer in the ISIS-I Satellite." Reviews of Scientific Instruments, 41, 1393-1402.

Heikkila, W. J., and J. D. Winningham, 1971. "Penetration of Magnetosheath Plasma to Low Altitudes through the Dayside Magnetospheric Cusps." To be published in the Journal of Geophysical Reasearch.

Johnson, F. S., 1965. Satellite Environment Handbook, Stanford University Press.

King, J. W., K. C. Reed, E. O. Olatunji and A. J. Legg, 1967. "The Behaviour of the Topside Ionosphere during Storm Conditions." Journal of Atmospheric and Terrestrial Physics, 29, 1355-1363.

McDiarmid, I. B., and J. R. Burrows, 1964. "Diurnal Intensity Variations in the Outer Radiation Zone at 1000 km." Canadian Journal of Physics, 42, 1135-1148.

Nelms, G. L., and G. E. K. Lockwood, 1967. "Early Results from the Topside Sounder in the Alouette II Satellite." Space Research VII, pp 604-623, North Holland Publishers, Amsterdam.

Peddington, J. H., 1960. "Geomagnetic Storm Theory." Journal of Geophysical Research, 65, 93-106.

Piddington, J. H., 1965. "The Morphology of Auroral Precipitation." Planetary and Space Science, 13, 565-577.

Rishbeth, H., and O. K. Garriott, 1969. Introduction to Ionospheric Physics, Academic Press, New York.

Tonks, L., and I. Langmuir, 1929. "Oscillations in Ionized Gases." Physical Research, 33, 195, 990.

Valley, S. L., 1965. Handbook of Geophysics and Space Environment, Air Force Cambridge Research Laboratory, McGraw Hill.

Van Allen, J. A., and L. A. Frank, 1959. "Radiation Around the Earth to a Radial Distance of 107,400 km." Nature, 183, 430-434.

Walt, W., and W. M. MacDonald, 1964. "The Influence of the Earth's Atmosphere on Geomagnetically Trapped Particles." Reviews of Geophysics, 2, 543-577.

CHAPTER 4

ATMOSPHERIC GRAVITY WAVES IN OUTLINE

Colin O. Hines

University of Toronto,
Toronto, Canada

4.1 General Nature

Atmospheric *gravity waves* are oscillations of the atmosphere whose nature is strongly affected by the action of
gravity. They occur with a wide range of periods and wavelengths, which is not amenable to precise specification but
which we may take as 10 minutes to 24 hours, and 100 meters
to 1000 km, respectively, in order to fix ideas temporarily.
In a sense, they include a part of the family of tidal oscillations (of period 24, 12, 8, ... hours) though the term
"gravity wave" is often employed in a more restricted sense
which will become apparent.

One aspect of gravity waves that is important in the
context of this outline is their tendency to grow in amplitude with height, which is a by-product of their tendency to
conserve energy flux despite an upward decrease in the density
of the gas that is supporting that flux (much as ocean waves
grow into breakers as they approach a sloping beach, in part
because the depth of water that is available to carry their
energy is decreasing). Gravity waves are of some importance
at low levels of the atmosphere, but because of this tendency
for growth they become of great importance at higher levels,
where in fact they constitute a dominant feature of the region's
dynamics.

4.2 Governing Equations for the Elementary Case

The elementary properties of gravity waves can be established best by considering the case defined by
(i) the equation of continuous mass conservation

$$\frac{\partial \rho}{\partial t} + \rho \, \nabla \cdot \mathbf{u} + (\mathbf{u} \cdot \nabla)\rho = 0, \qquad (4.1)$$

(ii)   the force equation

$$\rho(\frac{\partial \mathbf{u}}{\partial t} + \{\mathbf{u} \cdot \mathbf{\nabla}\}\mathbf{u}) = -\mathbf{\nabla}p + \rho\mathbf{g} \tag{4.2}$$

in which pressure gradients and gravity are the only operative
forces causing acceleration, and
(iii) an assumption of adiabaticity in the oscillation:

$$\rho(\frac{\partial p}{\partial t} + \mathbf{u} \cdot \mathbf{\nabla}p) = \gamma p(\frac{\partial \rho}{\partial t} + \mathbf{u} \cdot \mathbf{\nabla}\rho) \tag{4.3}$$

Here ρ, $\mathbf{u}$, p and γ are the atmospheric parameters density,
velocity, pressure and specific-heat ratio, respectively. The
latter is taken to be a property of the gas only, invariant
even in the presence of the waves ($\gamma=1.4$ for air). The others
are taken to be perturbed (by primed amounts) from certain
background values (subscripted *zero*): $\rho = \rho_0 + \rho'$, $\mathbf{u} = \mathbf{u}_0$
$+ \mathbf{u}'$, $p = p_0 + p'$, and for the present we shall assume that the
background wind velocity $\mathbf{u}_0$ vanishes. We may substitute these
forms into (4.1)-(4.3), and expand all products that appear,
and we then find that those equations contain some terms with
no perturbation (i.e., primed) parameters, some terms with a
single perturbation parameter, some with two, and some with
three. We now *linearize* the equations by ignoring all terms
that have more than one perturbation parameter. This is done
on the grounds that the perturbations are presumed to be *small*,
and products of perturbation quantities negligible - an assumption that can be justified (or not) a posteriori in application
to any particular system of waves. The result is

$$\frac{\partial \rho'}{\partial t} + \rho_0 \mathbf{\nabla} \cdot \mathbf{u}' + \mathbf{u}' \cdot \mathbf{\nabla}\rho_0 = 0 \tag{4.4}$$

$$\rho_0 \frac{\partial \mathbf{u}'}{\partial t} + \mathbf{\nabla}p' - \rho'\mathbf{g} = -\mathbf{\nabla}p_0 + \rho_0\mathbf{g} \tag{4.5}$$

$$\rho_0 (\frac{\partial p'}{\partial t} + \mathbf{u}' \cdot \mathbf{\nabla}p_0) = \gamma p_0 (\frac{\partial \rho'}{\partial t} + \mathbf{u}' \cdot \mathbf{\nabla}\rho_0). \tag{4.6}$$

We shall be searching for wavelike solutions for the perturbation quantities, and all terms on the left of (4.5) will
therefore be variable with time. We may conceive of a background state, which contributes the terms on the right of (4.5),
which does not vary with time (or, at any rate, varies on a
much longer time scale). The equality of the two sides of (4.5)
can then be obtained continuously as time varies, only if each
side separately vanishes. Thus,

$$\mathbf{\nabla}p_0 = \rho_0\mathbf{g}. \tag{4.7}$$

This is the hydrostatic relation for the background atmosphere. We will further simplify our system by assuming that the background state is one of constant temperature, in which p_0/ρ_0 must be a constant; say

$$p_0/\rho_0 = C^2/\gamma \qquad (4.8)$$

where C is a constant which will turn out to be the *speed of sound* appropriate to the background temperature. If we now choose Cartesian axes (x, y, z) such that the z axis is directed upward, opposite to **g**, (4.7) and (4.8) combine to yield

$$\frac{\partial p_0}{\partial z} = \frac{dp_0}{dz} = -\rho_0 g = -p_0/(C^2/\gamma g) = -p_0/H \qquad (4.9)$$

where H, the *scale height*, is defined by the equation itself, and is seen to be a constant under present assumptions. This permits immediate integration of (4.9), and with substitution in (4.8) as well we find

$$p_0 = p_g \exp - (z/H) \; ; \; \rho_0 = \rho_g \exp - (z/H) \qquad (4.10)$$

where p_g and ρ_g are certain constant *ground-level* values of p_0 and ρ_0, respectively.

When (4.7) is inserted in (4.5), and when equations (4.10) are employed to express $\nabla\rho_0$ and ∇p_0 in (4.4) and (4.6), the set (4.4)-(4.6) reduces to

$$\frac{\partial \rho'}{\partial t} + \rho_0 \, \nabla\cdot\mathbf{u}' - \rho_0 u'_z/H = 0 \qquad (4.11)$$

$$\rho_0 \frac{\partial \mathbf{u}'}{\partial t} + \nabla p' - \rho'\mathbf{g} = 0 \qquad (4.12)$$

$$\rho_0 (\frac{\partial p'}{\partial t} - p_0 u'_z/H) = \gamma p_0 (\frac{\partial \rho'}{\partial t} - \rho_0 u'_z/H) . \qquad (4.13)$$

These constitute the governing equations for the wave system in present circumstances, though their subsequent reduction makes use also of the definition (4.8) for C^2 and that in (4.9) for H.

4.3 Elementary Plane-Wave Solutions

One can proceed in various ways to search for solutions to (4.11)-(4.13), and those solutions may be of various forms (corresponding to plane waves, cylindrical waves, etc.), depending on the purpose in hand. For our purposes, it is best to discuss plane waves that propagate in a direction perpen-dicular to the y axis, say, for which $\partial/\partial = u_y' = 0$. It may be confirmed that the following are solutions of the type sought:

$$\frac{\rho'}{\rho_0 R} = \frac{p'}{p_0 P} = \frac{u_x'}{X} = \frac{u_z'}{Z} = A \exp\left(\frac{z}{2H}\right) \exp\{i(\omega t - k_x x - k_z z)\} \quad (4.14)$$

where A is an arbitrary amplitude factor, ω is an arbitrary (radian) frequency, k_x and k_z are the horizontal and vertical components of some *wave vector* **k** whose direction specifies the direction of phase propagation, and R, P, X and Z are certain constants which, when multiplied into the right-hand side of (4.14), yield the instantaneous amplitude of the partial density perturbation (ρ'/ρ_0), the partial pressure perturbation (p'/p_0), the horizontal wind perturbation u_x' and the vertical wind perturbation u_z', respectively. The use of complex solu-tions is of course an artificial expedient: the physical solutions are the real parts of the expressions given by (4.14) which are themselves automatically solutions of (4.11)-(4.13).

These are solutions, rather, under the prescription

$$k_z^2 = (\omega_g^2/\omega^2) k_x^2 - k_x^2 - \frac{1}{4H^2} + \frac{\omega^2}{C^2} \quad (4.15)$$

where

$$\omega_g^2 \equiv (\gamma - 1) g^2/C^2 \quad (4.16)$$

and under the further prescriptions

$$R \equiv \omega^2 k_z + i(\gamma-1)gk_x^2 - i\gamma g\omega^2/2C^2 \quad (4.17)$$

$$P \equiv \gamma\omega^2 \{k_z - i(1-\gamma/2)g/C^2\} \quad (4.18)$$

$$X \equiv \omega k_x C^2 \{k_z - i(1-\gamma/2)g/C^2\} \quad (4.19)$$

$$Z \equiv \omega\{ \omega^2 - k_x^2 C^2\}. \quad (4.20)$$

All of these derive simply from insertion of (4.14) in (4.11)-(4.13), except that the definition of any one of (4.17)-(4.20) is an arbitrary choice.

Examination of (4.15) shows that, between the frequency ω_g and the frequency

$$\omega_a \equiv \gamma g/2C \quad (4.21)$$

(which must be somewhat greater), one cannot have *both* k_x and k_z real: phases cannot be propagated in any direction, without

an exponential attenuation occurring in some other direction.
At frequencies greatly exceeding ω_a, (4.15) becomes to approximate to the simple sound-wave relation $k_x^2 + k_z^2 = \omega^2/C^2$, and
the waves may be termed *acoustic*; ω_a is the acoustic cut-off
frequency. At frequencies below ω_g, the effects of gravity
are important and the waves may be termed *gravity waves*; ω_g
is the gravity-wave cut-off frequency.

Important approximations arise in the gravity-wave case
when $|k_z| \gg \dfrac{1}{2H}$ and $\omega \ll \omega_g$. Then

$$k_z^2 \simeq (\omega_g^2/\omega^2)\ k_z^2 \qquad (4.22)$$

$$R \simeq i(\gamma-1)\ g\ k_x^2 \qquad (4.23)$$

$$P \simeq \gamma\omega^2 k_z \qquad (4.24)$$

$$X \simeq \omega k_x k_z C^2 \qquad (4.25)$$

$$Z \simeq -\omega k_x^2 C^2 \qquad (4.26)$$

These approximations apply to much of the observed gravity-wave
spectrum. From (4.22) we see that the angle of ascent of the
phases, α, is given by

$$\tan \alpha = k_z/k_x \simeq \omega_g/\omega = \tau/\tau_g \qquad (4.27)$$

where τ is the wave period $(=2\pi/\omega)$ and $\tau_g = 2\pi/\omega_g \ll \tau$ by pre-
sent assumptions. In these circumstances then, $|\tan \alpha|$ is
large and the phases propagate nearly vertically, upwards or
downwards; phase planes are nearly horizontal: spatial variations occur predominantly in the vertical direction, and only
to a lesser extent horizontally. From (4.25) and (4.26) we
see that

$$X/Z = -k_z/k_x \qquad (4.28)$$

which implies that the wind vector **u** is perpendicular to the
wave vector **k**: the winds are shearing winds, nearly horizontal and reversing in direction vertically every half wavelength.

The characteristics of winds revealed by *meteor trails*
and by rocket-released *vapor trails* in the upper atmosphere
are in accordance with these theoretical characteristics.
Their time scale of variation (corresponding to τ) is long in
comparison with τ_g ($\sim$ 5-10 minutes at the relevant heights);
they vary rapidly in the vertical (with equivalent $\lambda_z \sim$ 1-30 km,
typically) and only slowly in the horizontal; and they are

primarily horizontal. Further, the exponential growth with
height predicted in (4.14) is observed in general, with departures that can be explained by complications to be discussed.
The characteristics of many *moving ionospheric irregularities*
are similarly in accord with the theoretical characteristics,
although for them the more exact relations (4.15)-(4.20) must
often be employed and the *movement* of the irregularity is
often the movement of the phase surfaces (with horizontal and
vertical "trace" speeds ω/k_x and ω/k_z, respectively) *rather
than* the movement represented by the wind vector **u'**.

Because of the anisotropy introduced by gravity, wave
energy does not generally move in the same direction as do the
phases. Instead, it moves with the group velocity whose x
and z components are $\partial\omega/\partial k_x$ and $\partial\omega/\partial k_z$, respectively, the dif-
ferentiations being performed subject to (4.15) being main-
tained intact. Whenever $|k_z| >> 1/2H$, as before, the direction
of this group velocity is essentially perpendicular to the
direction of phase propagation, the horizontal components of
the two being in the same direction, but the vertical directions being opposed. Thus, waves whose energy propagates
obliquely upwards are subject to phase progression obliquely
downward.

One may expect that most sources of observed upper atmospheric gravity waves will originate in the energy-bearing
regions of lower levels. Energy would then be propagating
upward and phases downward, in the region of observation, and
this is in accord with the general run of observations.

4.4 Complications of Nonlinearity

The foregoing simple discussion must break down at some
height even with the simplest of atmospheric characteristics
assumed, for the exponential growth contained in (4.14) must
ultimately lead to amplitudes of such a large magnitude as to
invalidate the conversion of the basic equations (4.1)-(4.3)
into the perturbation equations (4.4)-(4.6). The nonlinear
terms (in primed quantities) can no longer be neglected,
once they become comparable to the linear terms: e.g., when
u'.$\nabla\rho'$ becomes comparable to $\partial\rho'/\partial t$, which is when **u'**·**k**$\approx\omega$ if
$|k| >> 1/2H$. As a first approximation, one might guess that
this is when $u_x' \sim \omega/k_x$ say. Observed values for ω/k_x are
typically 10-100 m/s, while wind speeds u_x' are of similar
amplitude at and immediately above meteor levels, so nonlinear effects must be anticipated. The situation is not
quite this serious, because of the previously noted tendency
for **u'** and **k** to be mutually perpendicular in a given wave, so

that $\mathbf{u'} \cdot \mathbf{k}$ tends to be much less than uk. When more than one
wave is present, however, the perpendicularity condition no
longer obtains: the $\mathbf{u'}$ of one wave need not be nearly perpendicular to the $\mathbf{k}$ of a second wave, and the condition for
nonlinear effects to be important is more readily established.
When nonlinearities are important, energy transfer from one
wave to another, the formation of new waves from mutual interactions of old waves, and the *breaking* of waves must all be
anticipated.

4.5 Complications of Instabilities

Quite apart from nonlinear effects in the wave equations,
the linear *perturbation* theory gives rise to temperature
gradients (with fractional variations of temperature given by
$T'/T_0 = p'/p_0 - \rho'/\rho_0$) which must become superadiabatic at
some levels as the exponential growth works its effects. The
atmosphere becomes convectively unstable in these conditions,
and *turbulence* may be expected to develop. This turbulence
will act to diffuse the energy and momentum from its organized
distribution in the wave system to a disorganized distribution,
and so converts wave energy first into turbulent energy and
then into heat. The situation may be more serious, in fact,
for turbulence might develop even without the vertical temperature gradient becoming superadiabatic. A *slantwise* super-
adiabicity is always possible, leading always to the possibility of instability, and wind shears in the wave system provide a further source of energy for driving turbulence. These
effects act to leech energy from the wave systems to heights
of 100 km or so, whereupon molecular effects appear to become
more significant as a dissipative mechanism.

4.6 Complications of Molecular Dissipation

Viscosity and *thermal conduction* are inherent properties
of the atmosphere, and their importance increases with height
as the mean free path increases. At some height they must
come to be significant in the propagation of waves, if the
waves have not already been removed by nonlinear or instability (or other) processes.

Properly, one should take molecular viscosity into account
by adding a force density $\mu(\nabla^2 \mathbf{u} + \nabla(\nabla \cdot \mathbf{u})/3)$ on the right-hand
side of (4.2) and so a corresponding primed quantity on the
right-hand side of (4.12), μ being the molecular viscosity.
Analytic solutions can no longer be obtained if one does this,
however, for the form of the equation is now altered: whereas other terms in (4.12) have the height variation $\exp(-z/2H)$,
this term has the height variation $\exp(+z/2H)$ since μ is

virtually height-independent. No adjustment to the form of solution can be found to permit accurate analytic corrections of a simple type. An elementary calculation, which is supported by more detailed analyses, does however permit an evaluation to be made of the circumstances under which viscosity becomes serious: we take it to be serious when the simple solutions (4.14), on insertion into the viscous force density term, yield a magnitude for that term that is comparable to the inertial force density. This yields the order-of-magnitude result (for $k_z \gg 1/2H$) $\omega \rho_0 u' \simeq \mu k^2 u'$ or

$$\omega/k^2 \simeq \mu/\rho_0 = \eta_M \qquad (4.29)$$

where η_M is the molecular kinematic viscosity. (If the atmosphere is turbulent, the eddy viscosity η_E replaces η_M in this criterion for important alterations to the wave system.) Equation (4.29) may be combined with (4.15), with $k_z \gg 1/2H$ assumed, to show that the largest k_z that is consistent with (4.29) is to be found at $\omega^2 = \omega_g^2/3$, when $k_z^2 \simeq 0.3 \, \omega_g/\eta_M$. (The precise numerical factor here depends upon the precise conditions assumed for viscosity becoming "important".) The corresponding vertical wavelength, $\lambda_z \simeq 2\pi \sqrt{3\eta_M/\omega_g}$, is of the order 1 km at a height of 100 km and increase upward. These wavelengths are indeed found to be *cut-off* wavelengths in the spectrum of observed zig-zag profiles of the wind at these heights, as revealed by rocket-released vapor trails; the height separations of successive *zigs* or *zags* almost invariably exceed these λ_z's.

Thermal conduction may be introduced properly only by rejecting the adiabatic equation (4.3), and inserting in its place the perfect-gas law plus an equation for heat transfer (including, in principle at least, a term that represents heat generation from viscous losses). Again, no analytic solutions are available in these circumstances. Because of the intimate relationship between thermal conduction and viscosity, however, both being a consequence of molecular transport processes, one might anticipate that thermal conduction would become important only when viscosity becomes important, and that its role as a dissipative mechanism would be quite analogous. These suspicions are borne out by detailed analysis.

4.7 Complications of Temperature Structure

The background temperature of the real atmosphere is not a constant, but rather varies substantially on a scale of a few kilometers in the vertical. One can infer the qualitative consequences of this by considering an elementary case in which one isothermal half-space is superimposed above another, of different temperature, there being a wave incident upon the interface from below. Certain interfacial conditions must be met: there must be no pressure discontinuity, and the vertical displacement must be the same on both sides of the interface, right at the interface itself. These conditions can be met continuously in time and in space along the interface only if a *transmitted* wave occurs in the upper half-space, with ω and k_x unchanged from the values obtaining in the incident wave. But (4.15) then implies that k_z must be different in the upper half-space, to make up for the differences in H and C: the transmitted wave is *refracted*, to propagate in some appropriate new direction.

Continuity of pressure and vertical displacement demand more, for they impose two conditions on the amplitudes of the wave systems. These conditions cannot be met simply by an appropriate change of A in (4.14), on going from the lower half-space to the upper, for that provides only a single degree of freedom (namely, the ratio of the two A's). Instead, it is necessary to add yet another wave, a *reflected* wave, propagating its energy away from the interface in the lower half-space.

The actual temperature profiles, being smoothly varying, give rise to continuous processes of refraction and internal reflection, with resultant interference effects that can be complex but must be treated in some detail to be treated at all adequately. About the only general statement that can be made is this: a given temperature structure will act to transmit or reflect waves with various ω's and k_x's in varying degree, and so will act as selective filter between sources in one region of the atmosphere and effects in another.

Beyond this, perhaps the simplest general point to be made is that the k_z deduced from (4.15) may be real for some heights and imaginary for others. The waves are "evanescent" when k_z is imaginary, and there is a strong tendency for the vertical flow of energy to be inhibited in such regions. Whether on this account or on others, strong reflection may occur, and the wave energy may be ducted between two heights of strong reflection (one of which may be the ground). Some energy may nevertheless escape above the duct region, and there the exponential growth with height could render its effects important despite its intrinsic weakness. This ducting effect may well account for an observed characteristic of many traveling ionospheric disturbances, their maintenance of strength over long distances of propagation: the main reservoir of

energy may well lie in a duct below 100-150 km, and may be
maintained with little loss despite the small leakage upward,
while the little that does leak upward may be observed in
the ionosphere and may appear to sustain no loss because it is
continually being replenished from below.

4.8 Complications of Wind Structure

The real atmosphere supports background winds, which have
been ignored so far in the preliminary analysis except for
their definition as $\mathbf{u_0}$. Examination of (4.1)-(4.3) will reveal
that these winds would alter the perturbation equations (4.11)-
(4.13) only by the addition of a $(\mathbf{u_0} \cdot \boldsymbol{\nabla})$ operation in conjunction with each $\partial/\partial t$ operation. In application to waves of
the form (4.14), this would result in each appearance of an ω
being replaced by the appearance of an 'intrinsic frequency'

$$\omega' \equiv \omega - \mathbf{u_0} \cdot (\mathbf{k} + i\mathbf{z}/2H) \tag{4.30}$$

upon insertion of (4.14) into (4.11)-(4.13), where $\mathbf{z}$ is a unit
vector in the z direction. If, as is usually assumed and as
is indeed the case in practice, the vertical component of $\mathbf{u_0}$
may be taken to vanish, (4.30) reduces to the elementary form
of a Doppler-shifted frequency,

$$\omega' = \omega - \mathbf{u_0} \cdot \mathbf{k} = \omega - u_{0x} k_x . \tag{4.31}$$

It is this frequency that must now be entered into formulae
(4.15), (4.17)-(4.20), (4.22)-(4.27).

As in the case of temperature structure, wind structure
imposes refraction and reflection, and opens the possibility of
wave ducts. It serves, more generally, as a selective filter
once again. The filter in this case has the further interesting property of being anisotropic: since the value of ω'
specified by (4.31) is dependent on the component of $\mathbf{u_0}$ in the
direction of wave propagation, the filtering effect will vary
with that direction. In this respect, winds add a dimension
to the filtering effect of temperature structure.

From (4.31) it will be apparent that ω' reduces to zero
whenever the horizontal trace speed ω/k_x matches the background
wind speed in the direction of propagation, u_{0x}. Any level at
which this occurs in the atmosphere is termed a 'critical
level', and very important processes occur at such levels, not
all of them understood. It seems likely that, for all practical
purposes, the energy of a wave is completely destroyed when it
reaches such a level, being converted to heat and to kinetic
energy of the background flow.

4.9 Complications of Earth Curvature and Rotation; Tides

The analysis to this point has taken gravity to lie in a single Cartesian direction, thereby ignoring the sphericity of the earth, and has made no provision for the inclusion of Coriolis and centrifugal forces associated with the earth's rotation. We now consider the complications introduced by these factors and so go beyond the topic of 'gravity waves' as defined in its more restrictive sense by the subject matter of the preceding sections. More specifically, we touch upon the tidal oscillations of the atmosphere, in par-ticular the tides with periods 12 hours and 24 hours.

At such long periods, the horizontal component of the Coriolis force density, $2\rho\mathbf{u}\times\mathbf{\Omega}$, must be included on the right hand side of (4.2) if the coordinate system is taken to rotate with the earth. The vertical component of this force, and the centrifugal force, are both ignored in tidal theory; and indeed, vertical accelerations are taken to vanish, which is consistent with the generally small value of ω that charac-terizes tides and with the small vertical velocities (relative to horizontal velocities) that might be inferred from a simple extrapolation of gravity-wave theory to such small ω's.

The sphericity of the earth is handled conveniently with the aid of spherical coordinates, r, θ,ϕ say, but it poses a further problem when contrasted with the circumstances of elementary gravity waves: the solutions must be compatible with the periodicity of 2π in the longitudinal coordinate ϕ, and they must be continuous at the poles $\theta=0$, π. The search for valid solutions then becomes a problem in eigenfunctions and eigenvalues.

The eigenfunctions contain longitudinal and latitudinal factors. The former are of the simple form $\exp(im\phi)$, where $m = 1$, 2, ... corresponding to 1, 2 ... wavelengths around the equator, and so to tides of period 24, 12 ... hours. The latter are known as Hough functions, which we may denote as $\theta_{m,n}$. The subscript m is identical to the m of the associated longitudinal function. The subscript n is a measure of the degree of complication in the north-south structure of the particular Hough function; for $m \geqslant 2$, n-m equals the number of nodal surfaces that lie between the north and south poles. The Hough functions, multiplied by their associated $\exp(im\phi)$ functions, are intimately related to spherical harmonies and may be expanded in terms of them. For each eigenfunction there is a corresponding eigenvalue, denoted $h_{m,n}$ which happens to have the dimensions of a length; it is known as the "depth of the equivalent ocean."

Given a certain 'm,n' mode of horizontal structure, the vertical structure of the tidal oscillation is associated with a vertical wave number k_z given by

$$k_z^2 = \frac{1}{Hh} \left[\frac{\gamma-1}{\gamma} + \frac{dH}{dr} \right] - \frac{1}{4H^2} \qquad (4.32)$$

This is analogous to (4.15) subject to certain caveats: the ω^2/C^2 term of (4.15) disappears because, with $\omega \ll \omega_a$, it is much smaller in magnitude than the $1/4H^2$ term; similarly the term $-k_x^2$ is ignored relative to the $(\omega_g^2/\omega 2)k_x^2$ term; the $(\gamma-1)$ of ω_g^2 (as defined by (4.16)) is replaced by $(\gamma-1) + \gamma\, dH/dr$, which serves to take some account of height variations of temperature; and the parameter h now plays the role previously played by ω^2/gk_x^2 (or, as one might say, the horizontal trace speed ω/k_x has been replaced by $\sqrt{gh}$, which is well known as the speed of a long wave in an ocean of depth h).

4.10 The Semidiurnal Tide

The semidiurnal tide consists of a superposition of oscillations of the '2,n' type. It is excited primarily by the absorption of solar radiation by ozone, at heights of 25-55 km say. The latitudinal distribution of the resultant heating most nearly resembles the $\theta_{2,2}$ function, and it leads then to a strong component of the '2,2' type. For this component, $h_{2,2} = 7.9$ km. It happens that the k_z which now results from (4.32) is imaginary in the mesosphere, at heights of 60-85 km roughly, and the energy of the '2,2' mode is strongly reflected there. Some transmission does occur, of a strength sufficient to render this mode an important one at meteor heights and above; but it has not the dominance at those heights that it enjoys at lower levels.

The '2,4' and '2,6' modes are excited less strongly, but the h's for them are smaller and the k_z's nowhere become imaginary. Their energy reaches meteor heights and overlying levels without suffering serious reflection, in consequence, and they themselves are relatively important in observations made at such heights.

4.11 The Diurnal Tide

The diurnal tide consists of a similar superposition. The '1,1' mode behaves in quite a different fashion from the

'2,2' mode, however, for it is confined primarily to latitudes below 50° and within those latitudes contains a node on each side of the equator. For it, $h_{1,1}$ = 0.7 km and this leads to vertical wavelengths (=$2\pi/k_z$) of about 25 km. Thus the '1,1' mode reverses in sign both in latitude and in height within dimensions over which the ozone heating remains strong and of a single sign; it is excited relatively inefficiently, in consequence. In fact, absorption of water vapor low in the atmosphere, over a height range small in comparison with the vertical wavelength, appears to be the dominant mechanism of excitation for this mode.

Despite its relatively modest means of excitation, the '1,1' mode is free to propagate through the mesosphere without serious reflection; it reaches meteor heights in considerable strength, giving rise to winds of the order 40 m/s in its low-latitude belt, and then rapidly dissipating itself through turbulent and molecular loss processes.

A second family of diurnal tidal modes exists, for which h is intrinsically negative; it is generally denoted by means of negative values for n. These modes of oscillation are of a different category from those so far discussed, being not of the general 'gravity wave' class. By virtue of their negative h's, their k_z's (which are still given by (4.32)) are intrinsically imaginary: energy deposited at some altitude in their excitation tends to remain at that altitude, without suffering vertical propagation away. The '1,-1' mode extends from pole to pole without a node, and closely matches the latitudinal variation of the ozone heating function. It is excited strongly in consequence, but appears strong only in the vicinity of the ozone layer. Higher-order 'negative' modes are also excited, primarily at higher latitudes (thus complementing the low-latitude bias of the 'positive' modes); but again, their energy remains near the level of excitation. The negative modes are of little concern at meteor heights, but local sources at somewhat greater heights may make them relevant there. Indeed, there is some evidence that the '1,1' mode, excited near heights of 100-130 km, may be the dominant source of the dynamo winds that lead to diurnal variations of ionospheric current systems and so of the ground-level magnetic variations that these systems produce.

4.12 Bibliography

Bretherton, F. P., 1966. "The Propagation of Groups of Internal Gravity Waves in Shear Flow." Quarterly Journal of the Royal Meteorological Society, 92, 466.

Butler, S. T., and K. A. Small, 1963. "The Excitation of Atmospheric Oscillations." Proceedings of the Royal Society, A274, 91

Eckart, C., 1960. Hydrodynamics of Oceans and Atmospheres, Pergamon Press.

Einaudi, F., 1970. "Shock Formation in Acoustic-gravity Waves." Journal of Geophysical Research, 75, 193.

Friedman, J. P., 1966. "Propagation of Internal Gravity Waves in a Thermally Stratified Atmosphere." Journal of Geophysical Research, 71, 1033.

Hines, C. O., 1960. "Internal Atmospheric Gravity Waves at Ionospheric Heights." Canadian Journal of Physics, 38, 1441.

Hines, C. O., 1963. "The Upper Atmosphere in Motion." Quarterly Journal of the Royal Meteorological Society, 89, 1.

Hines, C. O., 1968. "Tidal Oscillations, Shorter-period Gravity Waves and Shear Waves." Meteorological Monographs, 9, 114.

Hines, C. O., and C. A. Readdy, 1967. "On the Propagation of Atmospheric Gravity Waves through Regions of Wind Shear." Canadian Journal of Physics, 72, 1015.

Hodges, R. R., 1967. "Generation of Turbulence in the Upper Atmosphere by Internal Gravity Waves." Journal of Geophysical Research, 72, 3455.

Hough, S. S., 1898. "On the Application of Harmonic Analysis to the Dynamical Theory of the Tides." Philosophical Transactions of the Royal Society, A189, 201 and A191, 139.

Lindzen, R. S., 1967. "Thermally Driven Diurnal Tide in the Atmosphere." Quarterly Journal of the Royal Meteorological Society, 93, 18.

Midgley, J. E., and H. B. Liemohn, 1966. "Gravity Waves in a Realistic Atmosphere." Journal of Geophysical Research, 71, 3729.

Pierce, A. D., 1965. "Propagation of Acoustic-gravity Waves in a Temperature- and wind-stratified Atmosphere." Journal of the Acoutical Society of America, 37, 218.

Pitteway, M. L. V., and C. O. Hines, 1963. "The Viscous Damping of Atmospheric Gravity Waves." Canadian Journal of Physics, 41, 1935.

Siebert, M., 1961. "Atmospheric Tides." Advances in Geophysics, 7, 105.

Tolstoy, I., 1963. "The Theory of Waves in Stratified Fluids, Including the Effects of Gravity and Rotation." Reviews of Modern Physics, 35, 207.

Wilkes, M. V., 1949. Oscillations of the Earth's Atmosphere, Cambridge University Press.

CHAPTER 5

NOCTILUCENT CLOUDS -
THEIR CHARACTERISTICS AND INTERPRETATION

Benson Fogle

National Center for Atmospheric Research,
Boulder, Colorado

5.1 Introduction

Noctilucent clouds (NLC), the highest known clouds of the planet Earth, are formed at the mesopause at an altitude of around 82 km mainly during summer months in high latitudes in both hemispheres. These clouds, observable only during twilight when the sun is between 6° and 16° below the horizon, are silvery in color and often resemble cirrus or cirrostratus with marked wave structure (Figs. 5.1 and 5.2). They often are of small area and very thin and of low brightness so that stars shine through them almost undimmed, but in both brightness and spatial extent they have a considerable range. NLC began to attract the attention of scientists in 1885; the first recorded observation, recognizing that they were an unusual and remarkable phenomenon, was made by T. W. Backhouse (1885) from Germany on June 8 of that year. Many other observations were made at about that time in the USSR and Europe; in later years the number reported rose and fell, partly depending on the attention given to them by interested observers and partly due to a real fluctuation in NLC activity. Beginning with the IQSY in 1964, a successful effort to put visual observations of NLC on a systematic basis was initiated so that routine observations of these clouds are now being made on many of the meteorological stations in the northern and southern hemispheres between 45° and 90° latitude. These stations use the standardized observing and reporting procedures outlined in the International Noctilucent Cloud (NLC) Observation Manual (1970), and the data from these stations are collected, analyzed and published by three centers-- Edinburgh, Scotland; Tartu, Estonia, and Toronto, Ontario. Over a thousand occurrences of NLC are now on record, and while much has been learned about their characteristics, many uncertainties remain especially as regards the composition of the cloud particles and how and why the clouds form.

Fig. 5.1 Noctilucent clouds observed at Watson, Lake, Yukon Territory on 26-27 July 1965.

Fig. 5.2 Noctilucent clouds observed at Grande Prairie, Alberta on 15-16 July 1965.

It is scarcely possible now to judge whether the observation of NLC will ever be of value in weather forecasting,
but as the interpretation of NLC becomes more certain and
better developed, they can certainly contribute to our knowledge and understanding of the general dynamics and thermo-
dynamics of the atmosphere. These clouds serve as a tracer of
atmospheric motions at the mesopause, a region about which
very little is known, and they provide direct evidence of wave
motions in the high atmosphere. Their occurrence in certain
months and over a certain range of latitude is a pointer to
large scale synoptic properties of the general circulation,
and the particular dates and particular places in high latitudes where they are visible point to local nonuniformities
in the circulation and conditions in those mesospheric regions.

5.2   The Height

Over the past 85 years, hundreds of height determinations
have been made (Jesse, 1896; Störmer, 1935; Witt, 1962; Burov,
1966). The values range from 74 to 92 km, the average being
around 82 km, the height of the mesopause, but part of the
range must be due to errors of observation (Fogle and
Haurwitz, 1966). Most of these height determinations are of
the lower border of the NLC layer which is sharply defined and
much brighter than the faint and diffuse upper border. Evidence of the occasional occurrence of two layers of NLC simultaneously has been reported by Störmer (1933), Grishin (1967)
and Witt (1962). This, if true, could be due to a double
mesopause whose existence has been suggested by Schilling
(1965) and Novozhilov (1967). According to Grishin (1967) and
Witt (1962), when two layers of NLC occur, the lower one consists of a thin structureless veil, while the upper one contains the well-defined NLC forms.

Most of the NLC height determinations have been made by
ground-based parallactic photography, but a few were made by
rocket-borne photometric measurements, and optical radar may
prove useful for future NLC height and thickness measurements.

5.3   Latitudes of Observation

In the Northern Hemisphere, the lowest and highest latitude sightings of NLC were made from 45°N and 80°N, respectively, but most of the observations have been made over the
latitude range of 50°N to 65°N. The optimum latitude for
viewing NLC is around 58° (Fogle, 1968; Pavlova, 1960). This
zone of NLC occurrence seems to be symmetric about the geographic pole rather than the geomagnetic pole.

In the Southern Hemisphere, the latitudes for observing NLC appear to be similar to that in the Northern Hemisphere. Reliable reports of Southern Hemisphere NLC have been made over the latitude range of 51°S to 70°S (Fogle and Haurwitz, 1966), but they are far fewer in number than for the Northern Hemisphere, owing to the relative scarcity of observers and the poor observing conditions in the 50°S - 70°S zone.

No reliable reports of naturally occurring NLC from places below 45° have been made, and it seems likely that NLC do not extend to such low latitudes. Whether or not NLC occur at latitudes polewards of 80° is difficult to establish from the kind of observations now available, since they are restricted to twilight which is unavailable in high latitudes during summer.

There is some evidence for a poleward recession of NLC in late summer (Paton, 1967; Fogle, 1968), but more data from the network of stations is needed before this can be established with certainty. Complicating the determination of the variation of NLC activity with latitude is the marked variation of observing time with latitude.

5.4   Seasonal Distribution

In the Northern Hemisphere, NLC have been reported as early as March and as late as October, but over 90% of the sightings are made during the summer months of June-August. The frequency distribution of recorded NLC sightings as a function of season shows that the majority of sightings were made after the summer solstice with the peak of activity occurring two to four weeks after the summer solstice (Fogle and Echols, 1965; Fogle, 1968).

Based on the limited available data (around twelve reliable reports), Southern Hemisphere NLC also appear to occur predominantly during the (austral) summer months of December, January, and February with the peak of activity occurring in early January, shortly after the austral summer solstice (Fogle, 1966).

5.5   Spatial Extent

Sometimes the whole boundary of a display of NLC can be seen and its area estimated; this may be only a few tens of thousands of km^2. The extent of sky at the level of 82 km visible from the ground is a circular area about 1000 km in radius and of area about 3 million km^2; but only a part of this can be illuminated by sunlight against a dark sky. One

NLC display widely observed over Alaska and Canada during
August 1963 was too extensive to be all visible from any one
place; combination of data from many observers indicated that
its area was more than 2.5 million km^2 (Fogle, 1966). Seldom,
however, do NLC displays exist in continuous sheets over such
wide areas. In general, during NLC displays, there are areas
of clear sky separating discrete patches of NLC, and even in a
given patch there is often a considerable range of brightness.
It is possible that NLC might at times be circumpolar and have
an area of order 10^8 km^2, but this if true would be difficult
to establish by observations of the kind so far available,
since NLC cannot be seen from the ground at all longitudes
simultaneously. Spatial extent studies could be extended by
satellite observations from an appropriate orbit.

5.6 Duration

Since NLC cannot be seen during the day when the sky
background is too bright or at night when the mesopause is not
sunlit, a determination of their maximum duration cannot be
made except in those few cases when they form and decay during
twilight.

NLC displays are usually quite persistent and their observed duration is usually one or more hours, but individual
parts of the clouds, particularly the billow structure, sometimes form and decay within ten minutes. Weak displays of
limited spatial extent also tend to be short-lived and last
for less than an hour (Fogle, 1966).

Occasionally, very bright NLC displays have been observed
during the entire twilight observing period at a particular
station and durations of 4-5 hours were recorded. The appearance and disappearance of these displays, which were
already well formed when first sighted, seemed to be controlled
by the brightness of the sky background rather than their
formation and decay at those times (Fogle, 1968). This would
suggest that the clouds can exist during the day, and that
durations of tens of hours may be possible. The occasional
occurrence of extensive NLC displays over the same location on
two or more successive nights (Fogle, 1968) tends to support
this contention, but observations of a different kind than now
available are required to determine the validity of this view.

5.7 Drift Motion

NLC velocity measurements show that, in the Northern
Hemisphere, NLC displays as a whole generally drift toward the

southwest at average speeds of 40 m/sec, but some of the wave
forms in NLC often move at speeds and in directions different
from the whole display (Witt, 1962; Fogle, 1966). Studies of
the limited Southern Hemisphere data show that NLC displays
there generally move toward the northwest at speeds similar
to that in the Northern Hemisphere (Fogle, 1966).

5.8 Wave Structure

The marked wave structure often observed in NLC provided
the earliest evidence of wave motions in the upper atmosphere.
The wave forms which occur in NLC have been classified into
two groups--bands and billows (WMO, 1970). The available
meager data on the wave parameters for bands and billows
(Haurwitz and Fogle, 1969) indicate that they have different
characteristics, although they may be produced by the same
mechanism. The short, closely spaced striations called billows are characterized by wavelengths of 3-10 km, crest widths
of 10-40 km, wave amplitudes of 0.5-1.0 km, lifetimes of 6-24
minutes and speeds of 50-100 m/sec. The longer, more widely
spaced striations called bands are characterized by wavelengths
of 10-75 km, crest widths of several hundred km, wave amplitudes of 1.5-3.0 km, lifetimes of 1-3 hours, and speeds that
differ from the NLC systems. Internal gravity waves seem to
be the likely cause of the NLC bands and possibly the billows
as well, but there is a possibility that the NLC billows are
Helmholz interface waves (Haurwitz and Fogle, 1969).

5.9 Thickness and Vertical Wave Amplitude

NLC are so tenuous that it is not easy to measure the
thickness, or even the vertical wave amplitude, which may be
greater. Witt (1962) has made measurements that suggest thickness of 2 km or less, and vertical wave amplitudes associated
with the band and billow forms of 0.5 to 3 km.

5.10 Diurnal Variation

Since NLC can be observed only during twilight hours, a
study of the diurnal variation of NLC activity must be restricted to a comparison of their activity during evening and
morning twilights. Such a study by Fogle (1968) showed that
during years of enhanced NLC activity (1964-66), NLC generally
first appear before midnight during evening twilight, that they
usually last on into morning twilight, and that they are
brighter and more extensive during morning twilight. These
results are in accord with those of Jesse (1896) and others
for the high NLC activity years of 1885-1887, but they do not

apply to years of low NLC activity. During years of low activity, the NLC are predominantly seen only during morning twilight (Jesse, 1890).

5.11   Auroral Influence

Paton (1954) was first to suggest that aurora can adversely affect NLC. In his observations of a simultaneous display of aurora and NLC on the night of 24 July 1950, he found that considerable turbulence developed in the NLC within an hour or two after onset of the aurora. He suggested that this turbulence was caused by considerable vertical motion induced at the mesopause in some way by the aurora.

An earlier observation of aurora and NLC was made by Vestine (1934). Although he did not comment on a possible relationship between aurora and NLC, in his description of the event, he mentioned that the intensity and extent of the NLC decreased for approximately half an hour after onset of the aurora, which lasted for five minutes. During an NLC field expedition to Canada in 1965, Fogle (1966) observed six simultaneous displays of aurora and NLC in which unusual changes occurred in the NLC after onset of the aurora. The general effects observed were a decrease in the area and intensity of the NLC (on two occasions they vanished completely) and a decay of their well-ordered structure. Ferry (1970) also reports apparent auroral influence on NLC. Other evidence pointing to auroral induced effects near the mesopause are Murcray's (1957) observation that the 9.6 micron band of O_3 may be enhanced by aurora, and Brown's (1970) theoretical results indicating that there are fewer water cluster ions near the mesopause after aurora than before. These data, while suggestive, are not conclusive and further observations and measurements are required to determine whether this is a real effect.

5.12   Year to Year Variation

Analysis of NLC data from 1885-1970 shows seven periods of enhanced activity of one to six years duration (see Table 5.1 During such periods, NLC were generally brighter, more extensive, of longer duration, and much more frequent than during quiescent periods. These periods of enhancement show no relationship with solar, cometary, or meteor activity, but each seems to have been preceded by an explosive volcanic eruption of the Plinian type, generally in equatorial latitudes, and/or multimegation nuclear weapons tests in the troposphere, both of which inject or entrain large quantities of water vapor into the stratosphere. The two periods of greatest

Table 5.1 Periods of enhanced NLC and associated events

Period of Enhanced NLC	Volcanic Eruptions or Nuclear Weapons Tests
1885-1889	Krakatoa, 1883
1908	Shtyubelya, Sapka, 1907
1911	Taal, 1911
1925-1929	Visokoi Isl., 1922-29; Gunung Batur, 1925-26
1932-1938	Quizapu, 1932; Tzerimai, 1937-38; Raluan, 1937; Ghai, 1937
1952-1960	Mt. Larrington, 1951; Bezymjannaya, 1955; Manam, 1956; Nuclear Weapons Tests, 1952, 1954-58
1963-1968	Agung, 1963; Taal, 1965; AWU, 1966; Nuclear Weapons Tests, 1961-62

enhancement of NLC activity followed the eruption of Krakatoa
(1883) and Agung (1963), both of which are near the equator.
These eruptions were, respectively, the largest and second
largest volcanic eruptions of the past 100 years.

5.13 Particle Size and Number Density

Polarization Measurements (Witt, 1960; Willmann, 1962)
have shown that the average diameter of NLC particles is about
0.3 micron and that the particles are nonmetallic and could
be stony meteoric dust or water substance or a combination of
the two. The number density of NLC particles has been estimated by Ludlam (1957) to be between 10^{-2} and 1 per cc.

5.14 Theory

A satisfactory theory of NLC must explain why they occur
at or near the mesopause, mainly in summer and in latitudes
above 45°, how they are formed, why they are present on some
nights but not on others, why they are thin, and why they
generally last for some hours.

Many speculations have been made as to the nature and
cause of NLC. For a long time, dust was believed to play an
important role in their formation. The early views that they
were volcanic dust clouds (Jesse, 1888) or cosmic dust clouds
(Vestine, 1934; Ludlam, 1957) were rejected in favor of their
being ice clouds (Humphreys, 1933; Hesstvedt, 1961, 1962;
Chapman and Kendall, 1965). A long series of temperature
measurements have shown that the temperature at the high
latitude summer mesopause is very low, 140°K or lower (Theon,
et al., 1967), and theoretical considerations (Hesstvedt,
1969) show that the water vapor pressure at the mesopause is
probably on the average 1-5 x 10^{-6} g/g, corresponding to a
frostpoint of around 148°K. Therefore, supersaturation is
likely to occur in a thin layer near the high latitude summer
mesopause, and nucleation of the ice phase could then take
place, followed by a growth of the particles.

Until recently it was generally believed that the
nucleation took place on cosmic dust particles; this view was
strengthened by the results of the first particle sampling
experiments (Witt, et al., 1964) which indicated dust particle
concentrations sufficiently high to form a visible cloud.
Subsequent sampling experiments, however, did not confirm the
first results and have indicated that there is insufficient
dust at NLC heights for dust to play an active role in their
formation (Soberman, et al., 1969; Witt, 1969; Frank, et al.,
1970). Other candidates for NLC sublimation nuclei, now being
considered, are water cluster ions such as $Fe(H_2O)_6^{2+}$ and
$H_3O^+(H_2O)_n$ (Hesstvedt, 1969; Witt, 1969), but the growth of
these ions to the particle sizes observed in NLC is a problem
which has not been explained.

Considerations of the mass budget of NLC and the fall speed of their particles show that, in the absence of updrafts, most of the water would be removed from the NLC layer in about one hour or less, a much shorter lifetime than has been observed for NLC. To reconcile this disparity, some mechanism is required to replenish the water vapor supply in the NLC layer. The most likely way for this to be done is by updrafts of moist air from below the mesopause of order 10 cm/sec (Charlson, 1965; Webb, 1965; Hesstvedt, 1969). This would replenish the water supply in the cloud layer, lower the temperature and impede the downward flux of particles, causing them to have a longer residence time and attain a larger size in the NLC layer.

5.15   References

Backhouse, T. W., 1885. "The Luminous Cirrus Clouds of June and July." Meteorological Magazine, Vol. 20, 133.

Brown, R. R., 1970. "On the Influence of Energetic Electron Precipitation on the Water Cluster Ion Population in the Upper D-Region." Journal of Atmospheric and Terrestrial Physics, Vol. 32, 1747-1753.

Burov, M. I., 1966. "Opredelenie Koordinat Serebristykh Oblahov." Meteorol. Issled. No. 12, II Razdel Programmi MMG, Izdat. "Nanka", Moscow, 33-46.

Chapman, S. and P. C. Kendall., 1965. "Noctilucent Clouds and Thermospheric Dust; their Diffusion and Height Distribution." Quarterly Journal of the Royal Meteorological Society, Vol. 91, 115-131.

Charlson, R. J., 1966. "A Simple Noctilucent Cloud Model." Tellus, Vol. 18, 451-456.

Ferry, Guy V., 1970. "Noctilucent Cloud Observation in Connection with a Sounding Rocket Launch." Journal of Geophysical Research, Vol. 75, 6857-6861.

Fogle, B. and C. Echols, 1965. "Summary of Noctilucent Cloud Reports from 1885-1964." Geophysical Institute Report No. UAGR-163, University of Alaska.

Fogle, B., 1966. "Noctilucent Clouds." Geophysical Institute Report No. UAGR-177, University of Alaska.

Fogle, B. and B. Haurwitz, 1966. "Noctilucent Clouds." Space Science Review, Vol. 6, 279-340.

Fogle, B., 1968. "The Climatology of Noctilucent Clouds According to Observations made from North America during 1964-1966." Meteorological Magazine, Vol. 97, 193-204.

Frank, E. R., J. P. Lodge, J. P. Shedlovsky and B. Fogle, 1970. "Particles Collected during a Noctilucent Cloud Display." Journal of Geophysical Research, Vol. 75, 6853-6856.

Grishin, N. I., 1967. "Dynamic Morphology of Noctilucent Clouds." Proceedings of an International Symposium on Noctilucent Clouds (Tallinn, 1966), Moscow, 1967, 193-199.

Haurwitz, B. and B. Fogle, 1969. "Wave Forms in Noctilucent Clouds." Deep Sea Research, Supplement to Vol. 16, 85-95.

Hesstvedt, E., 1961. "Note on the Nature of Noctilucent Clouds." Journal of Geophysical Research, Vol. 66, 1985-1987.

Hesstvedt, E., 1962. "On the Possibility of Ice Cloud Formation at the Mesopause." Tellus, Vol. 14, 290-306.

Hesstvedt, E., 1969. "Nucleation and Growth of Noctilucent Cloud Particles." Space Research IX, 170-174.

Humphreys, W. J., 1933. "Nacreous and Noctilucent Clouds." Monthly Weather Review, Vol. 61, 228

Jesse, O., 1888. "Ueber die Leuchtenden (Silbernen) Wolken." Meteorologische Zeitschrift, Vol. 5, 90-94.

Jesse, O., 1890. "Luminous Clouds." Nature, Vol. 43, 59-61.

Jesse, O., 1896. "Die Höhe der Leuchtenden Nachtwolken." Astr. Nachrichten, Vol. 410. 161.

Ludlam, F. H., 1957. "Noctilucent Clouds." Tellus, Vol. 9, 341-364.

Murcray, W. B., 1957. "A Possible Auroral Enhancement of Infrared Radiation Emitted by Atmospheric Ozone." Nature, Vol. 180, 139-140.

Narcisi, R. S., 1967. "Ion Composition of the Mesosphere." Space Research VII, 186-196.

Paton, J., 1954. "Direct Evidence of Vertical Motion in the Atmosphere at a Height of about 80 km Provided by Photographs of Noctilucent Clouds." Proceedings of the Toronto Meteorological Conference, 1953, 31-33.

Paton, J., 1967. "Noctilucent Clouds over Western Europe during 1967." Meteorological Magazine, Vol. 97, 174-176.

Pavlova, T. D., 1960. "Apparent Frequency of the Appearance of Noctilucent Clouds based on the Observations at the Stations of the Hydrometeorological Service for 1957 and 1958." Izd. IGU Astronomicheskaia Observatoria Issleedovania Serebristykh Oblakov, Vol. 1, 3-58.

Schilling, G. F., 1965. "Latitudinal Variation of Mesopause Height Inferred from Eclipse Observations." Journal of the Atmospheric Sciences, Vol. 22, 110-115.

Soberman, R. K., S. A. Chrest and R. F. Carnevale, 1969. "Rocket Sampling of Noctilucent Cloud Particles during 1964 and 1965." (In print).

Störmer, C., 1933. "Height and Velocity of Luminois Night Clouds Observed in Norway 1932." University of Oslo Publication 6.

Störmer, C., 1935. "Measurement of Luminous Night Clouds in Norway 1933 and 1934." Astrophysica Norwegica, Vol. 1, 87-114.

Theon, J. S., W. Nordberg and W. S. Smith, 1967. "Temperature Measurements in Noctilucent Clouds." Science, Vol. 157, 419-421.

Vestine, E. H., 1934. "Noctilucent Clouds." Journal of the Royal Astronomical Society of Canada, Vol. 28, 249-272, 303-317.

Webb, W., 1965. "Morphology of Noctilucent Clouds." Journal of Geophysical Research, Vol. 20, 4463-4474.

Willmann, Ch., 1962. "On the Polarization of Light from Noctilucent Clouds." (In Russian). Translations of the Conference on Noctilucent Clouds, Tallin, p. 29.

Witt, G., 1962. "Height Structure and Displacement of Noctilucent Clouds." Tellus, Vol. 14, 1-18.

Witt, G., C. L. Hemenway and R. K. Soberman, 1964. "Collection and Analysis of Particles from the Mesospause." Space Research IV, 197-204.

Witt, G., 1969. "The Nature of Noctilucent Clouds." Space Research IX, 157-169.

CHAPTER 6

NOCTILUCENT CLOUD WAVE STRUCTURE

B. Haurwitz

National Center for Atmospheric Research,*
Boulder, Colorado

6.1   Introduction

Most noctilucent cloud (NLC) displays show wave forms.
These waves are the earliest observed indications of *wave
motions* in the high atmosphere, and even today they represent
the only naturally occurring evidence of wave forms at these
levels.   Thus, they can serve as a valuable supplement to
such other data on wave motions in the upper mesosphere and
lower thermosphere, as are obtained from radio and visual
meteor trains, from artificial clouds, and from measurements
of ionospheric parameters.

The studies of NLC waves made so far have not had available the instrumentation required for a determination of all
the important parameters.   Hence the data which can be used
for theoretical studies are quite incomplete.   These data
and the   necessarily tentative   conclusions or speculations
based on this information will be discussed in the following
two sections.   In the last section a more complete observational
program will be outlined.

6.2   The Wave Observations

As in the case of clouds in the troposphere and lower
stratosphere, the NLC are classified in different *morphological*
groups (World Meteorol. Organization, 1970).   Of these groups
the longer bands and the shorter billows are of interest here
since they exhibit wave forms.   The results of the measurements of wave parameters have been compiled by Haurwitz and
Fogle (1969) and are summarized in Table 6.1.   It cannot be

*The National Center for Atmospheric Research is sponsored by
the National Science Foundation.

Table 6.1 Characteristic Parameters of NLC Waves

	Billows	Bands
Wavelength	3-10 km	10-75 km (or larger?)
Crest width	10-40 km	several 100 km
Wave amplitude	0.5-1 km	1.5-3 km
Lifetime	6-24 min	1-3 hrs
Speed	50-100 m/sec	diff. from NLC systems

emphasized too strongly that this table is based on a very
small number of data, 15 cases for the billows, 6 for the
bands. In no case were all the parameters measured. Thus
the lifetimes are only based on newer timelapse films taken
since 1964 by Fogle in North America, the amplitudes on
stereoscopic photographs by two ballistic cameras used by
Witt (1962) during an NLC display in Sweden. Nevertheless,
Table 6.1 attempts to summarize for the purpose of further
discussion what appear, in the twilight of our present knowledge, the characteristic parameters for *Billows* and *Bands*.

The bands are distinguished from the billows by their
larger dimensions, namely, greater wavelengths and longer
crests. It is not possible to deduce from the limited number
of data if there is really a distinct gap between these two
forms, or if a greater number of cases would show a gradual
transition between the two. Witt's (1962) amplitude determinations indicate that the bands have larger amplitudes than
the billows, as might be expected. The North American data
indicate that the lifetime of the bands is longer than that
of the billows, a fact which will be discussed in Section 6.3.
The speeds quoted in the table are measurements of displacements of NLC features. Because the wave and group velocities
are in general different from the speed of the air, it is
not *a priori* clear whether the *speed* in the table represents
an air motion or the rate of displacement of the wave forms.
In Section 6.3 this question will be discussed in more detail
from a theoretical viewpoint.

In the case of wave motion one would generally expect
the wave propagation to be in the direction *normal* to the
wave crests. In the case of the billows, the normal to the
wave crests coincides very nearly with the observed direction
in two cases, but it differs in the other case where both
parameters are known. For the bands only two simultaneous
measurements of crest normal and direction of motion exist.
In one case they coincide very nearly, in the other they differ by a substantial amount. Any conclusion concerning a
possible difference of billows and bands with respect to the
direction of motion and of crest normal will thus obviously

have to be postponed until more measurements have been accu-
mulated.

6.3 Theoretical Speculations

It is now generally assumed that NLC consist of *ice
crystals* condensed on some condensation nuclei or on hydrated
ions. With this assumption there is no difficulty to explain
the appearance of wave forms in NLC: Where the air ascends
because of the wave motion, condensation occurs because of the
adiabatic cooling while sublimation occurs in the regions of
descending air motion and adiabatic heating. Where condensation and sublimation are less intense, the NLC waves will appear
as arrays of brighter and less bright bands and billows rather
than as cloud wave crests and dark sky.

A necessary condition for this explanation of the NLC
wave forms is, of course, that the growth and decay of the
ice coating occurs fast enough to be effective during the
up- and downward motion. Estimates by Hesstvedt (1962) show
that this is the case at least for sublimation nuclei. For
hydrated ions whose possible role in the formation of NLC has
only recently been advocated a similar investigation has not
yet been made.

For the sake of completeness it may be pointed out that it
would also be possible to account for the appearance of wave
forms if the NLC particles are not composed of a volatile
substance. If an undulating NLC layer is viewed approximately
at right angles to the crests, brighter and darker bands will
be seen by the observer since his line of sight traverses alternately larger and smaller distances of this wavy layer.
In addition to this effect of the viewing angle the horizontal
divergence and convergence caused by the wave motion will give
rise to periodic changes of the total number of NLC particles
in vertical columns of unit cross sections, again resulting in
the appearance of darker and brighter bands or billows (Haurwitz, 1961).

The appearance of the NLC billows is very similar to that
of the tropospheric billow clouds, suggesting that they may
represent a similar physical process, namely, waves at an
interface where the background wind and temperature change
rapidly in vertical direction, as was first suggested by Helmholtz (1889). If it is assumed that the wave system moves with
the speed of the vector mean of the winds at the interface
(Wegener, 1911), the wavelength can be computed from the
wind and temperature discontinuities. In the few cases where
the required parameters in billow clouds have been measured,

good agreement has been obtained between observed and computed wavelength (Haurwitz, 1947). The few simultaneous observations of wind velocities and billow orientation show
further that the wave system moves with the speed of the mean
wind,so that Wegener's assumption is satisfied, and that the
crests are normal to the wind shear, as also assumed by
Wegener.

In the case of the NLC waves it is found that a wavelength of 10 km would require a wind discontinuity of 85 m
sec^{-1} if they are interface waves. This value may not be
excessively high, and some evidence of such wind shears has
been reported by Theon, Nordberg, Katchen, and Horvath (1967).
But the wind shear would have to be concentrated in a layer
whose thickness is small compared to the wavelength in order
to treat the motion as an interface wave; thus, at least the
interpretation of the longer billows as interface waves appears
doubtful. For bands with their longer wavelengths, the wind
discontinuities required, if they are interface waves, become
unrealistically large—for instance, 425 m sec^{-1} for waves of
50 km length.

It is therefore very likely that at least the bands are
not interface waves, but rather *internal gravity waves*. The
dispersion relation for these wave types gives horizontal trace
velocities of reasonable orders of magnitude, similar to those
observed in NLC displays. Because measurements of the required
meteorological parameters are still lacking, it is at present
only possible to make order-of-magnitude comparisons between
theory and observations. Nevertheless it seems likely in view
of the foregoing discussion that both billows and bands are
internal gravity waves. But a final decision, especially in
the case of billows, requires that we have more detailed information on the meteorological parameters at NLC levels.

That the lifetime of the shorter billows is measured in
minutes while that of the longer bands is measured in hours
may be due to viscous damping. The effect of (eddy or molecular) *viscosity* in the hydrodynamic equations is given by
terms such as

$$\nu \partial^2 u / \partial x^2$$

where ν is the kinematic coefficient of viscosity, u and x,
respectively, a horizontal velocity component and a horizontal
coordinate.

The wave factor of u has the form

$$\exp i(\omega t - k_x x - k_y y - k_z z)$$

where $k_x = 2\pi/L_x$ is the wave number component in the x direction, k_y and k_z are the wave number components in the other two coordinate directions. Hence

$$\nu \partial^2 u / \partial x^2 = -\nu 4\pi^2 u / L_x^2$$

showing that frictional dissipation is larger for shorter than for longer waves. Thus one would generally expect that the lifetime of the longer bands exceeds that of the shorter billows, as is observed. But this general conclusion applies not in each observed case. Such deviations are hardly surprising since the lifetime of the wave systems does depend not only on the rate of energy dissipation, but also on the rate at which energy is fed into the system.

As has already been pointed out in the preceding Section, it is not clear *a priori* if the observed rates of displacement are air speeds or phase or group velocities. Only measurements of the displacements of various features in an NLC display can give clues for a decision. Witt's (1962) detailed observations of the display of 10 August 1958 showed that the NLC system as a whole and the billows moved toward SW with speeds of 50 to 100 m sec^{-1}, while the bands were displaced to the NE at the rate of 10 to 15 m sec^{-1}. At the same time the air speed, as determined by bright patches in the display, was towards SW with speeds between 60 and 120 m sec^{-1}. Similarly, the time lapse films of the North American displays show that the billows move in general with the NLC system while the displacement of the bands is different from it.

This difference between the motions of the billows and the bands can be explained as a consequence of their different wavelengths. The horizontal trace velocity c_x of internal gravity waves is small for short wavelengths, but increases with the wavelength. If the ambient air is not at rest, but has a velocity U, for the sake of simplicity assumed constant here, c_x in the dispersion relation is replaced by $c_x - U$. This difference is small for the short billows, but large for the long-wave bands. Thus,

for the billows $c_x = U \pm$ small quantity

for the bands $c_x = U \pm$ large quantity

The two signs occur because the dispersion relation for the internal gravity waves and high-frequency acoustic-gravity

waves is biquadratic. Thus the billows will move with speeds
similar to those of the ambient air while the motion of the
bands may be very different from U and even in the opposite
direction.

Various speculations about the *sources of energy* of NLC
waves have been briefly enumerated by Haurwitz and Fogle (1969).
At present the most plausible hypothesis appears to be the one
advanced by Hines (1968), namely, that the required energy is
transmitted from the troposphere up to the NLC levels. Because
information on the meteorological parameters affecting this
energy propagation, namely, temperature and wind, is almost
completely lacking for cases of NLC waves, Hines' arguments
are necessarily very tentative, but he could at least demonstrate that his hypothesis is not in conflict with any known
facts. He was further able to show that in the lower troposphere enough energy is available in the relevant spectral
region to account for the (estimated) energy of the NLC bands.
The general possibility of the propagation of internal gravity
waves to NLC heights had also been demonstrated by Zhukova
and Trubnikov (1967).

6.4   Suggestions for Further Studies

1)   A prime requirement for any satisfactory study of
NLC wave forms is the availability of two ballistic cameras.
A number of the wave parameters of interest, such as the wavelength, crest width and orientation, speed and direction of
the waves, can be obtained from single-camera time-lapse pic-
tures if the NLC height is assumed to be known. While the
latter appears to lie within fairly narrow limits, the fact
that it cannot be measured by observations from one location
will affect the accuracy of the other measurements. More
important, two cameras permit stereoscopic viewing and measurements and thus a better resolution of details, including
the determination of wave amplitudes.

2)   To follow the development and disappearance of the
wave forms time-lapse films should be taken continually during
NLC displays.

3)   To collect information on NLC brightness, particle
size, and number density in different parts of the waves
spectro-photometric and polarization measurements of the
fine structure in NLC displays should be obtained. If the
visibility of the NLC depends on particle growth due to ice
accretion, as is now generally believed, then such data may
also allow estimates of the vertical component of the air
motion.

4) Since the wave motions depend on temperature, wind, and humidity, observations of these quantities, at least in some sample displays of NLC waves, are also needed, necessitating the development of suitable instrumentation for humidity determinations at these levels. Since the energy sources of the wave motions may be at least often, if not always, in the low atmosphere, a knowledge of temperature and wind distribution from there to the NLC level is required to determine whether the atmospheric conditions are such that the energy can propagate upward to the mesopause. Because of the obliqueness of the ray paths, the distribution of temperature and background wind will probably not be known in sufficient detail to establish such a path unambiguously, but the accuracy of any ray tracing could at least be substantially improved.

6.5 References

Haurwitz, B., 1947. "Internal waves in the atmosphere and convection patterns." Annals of the New York Academy of Sciences, Vol. 48, 727-744.

Haurwitz, B., 1961. "Wave formations in noctilucent clouds." Planetary and Space Sciences, Vol. 5, 92-98.

Haurwitz, B., and B. Fogle, 1969. "Wave forms in noctilucent clouds." Deep-Sea Research, Supp. to Vol. 16, 85-95.

Helmholtz, H. von, 1889. "Ueber atmosphaerische Bewegungen, II." Sitz.-Berlin Akademy Wisschaft, Berlin, 761-780.

Hesstvedt, E., 1962. "On the possibility of ice cloud formation at the mesopause." Tellus, Vol. 14, 290-296.

Hines, C. O., 1968. "A possible source of waves in noctilucent clouds." Journal of Atmospheric Sciences, Vol.25, 937-942.

Theon, J. S., W. Nordberg, L. B. Katchen, and J. J. Horvath, 1967. "Some observations on the thermal behavior of the mesosphere." Journal of Atmospheric Sciences, Vol. 24, 428-438.

Wegener, A., 1911. Thermodynamik der Atmosphäre. J. A. Barth, Leipzig.

Witt, G., 1962. "Height, structure and displacements of noctilucent clouds." Tellus, Vol. 14, 1-12.

World Meteorological Organization, 1970. International Noctilucent Cloud Observation Manual. No. 250. TP. 138.

Zhukova, L. P., and B. M. Trubnikov, 1967. "On the dynamics
of mesoscale wave movements in the field of noctilucent clouds
and the penetration of the tropospheric mesoscale waves into
the upper layers of the atmosphere." Proceedings of the
International Symposium on Noctilucent Clouds, Tallin, 1966,
Moscow, 1967, 222-230.

CHAPTER 7

METEOR TRAIL RADAR WINDS OVER EUROPE

Andre Spizzichino

Centre National D'Etudes des Telecommunications,
ISSY-LES-Moulineaux, France

7.1   Introduction

The first systematic series of wind measurements by
meteor radar observations were made at Stanford (U.S.A.) in
1950 (Manning et al., 1950). Wind values were obtained in the
80-105 km altitude range, where they were found to have strong
variations, but these variations could not be analyzed because
of the lack of height determination for each detected echo.

Some attempts were made later at Jodrell Bank (U.K.)
(Greenhow, 1954; Greenhow, Neufeld, 1955 a and b, 1956, 1961)
and at Adelaide (Australia) (Elford, 1959, 1964) to obtain
both wind measurements and a determination of the altitude
of the measurement. It was difficult to obtain both a high
hourly rate of data and an accurate altitude determination, and
it has not been possible to reconstitute the variation of the
wind profiles. Meteor wind measurements made before 1965 only
allowed an analysis of the wind components which depend only
slightly on altitude; i.e.,the *prevailing wind* and the *semi-
diurnal tide*.

The experiment described below is a first attempt to
obtain a continuous observation of the wind structure with
respect to height and time. A continuous wave radar has been
used,and its high sensitivity has enabled us to obtain a high
measuring rate and to determine the altitude within an accuracy
of $\pm$ 1 km.

The description of this equipment has already been given
elsewhere (Spizzichino et al., 1965; Revah, 1969) and will be
briefly mentioned in Section 7.2. A discussion of the calibration and the errors is given in Section 7.3. Section 7.4
deals with the hourly rate of obtained data and their comparison with that of other experiments.

The data processing, described in Section 7.5, brings to light a new and difficult problem concerning the interpolation to obtain the wind pattern and its harmonic analysis.

Section 7.6 gives some results of a first series of re-cording runs made in 1965-66 at Garchy (47°N, 3°E). These data were obtained under less satisfactory conditions than the present records (shorter recording runs, and the measurement of the zonal component of the wind only) and are then pre-sented as preliminary results. These data brought to light some new phenomena: propagation and partial reflections of *gravity waves*, propagation of the first mode of the *diurnal tide* with a perturbed phase, etc.

7.2 Description of the Equipment

The radar built by the Centre National D'Etudes des Telecommunications for meteor observations has been operating since 1965 and has been described in detail in previous papers (Spizzichino et al., 1965; Revah, 1968, 1969). It uses con-tinuous waves of a frequency of 29.8 MHz. The transmitting and receiving stations are located respectively at Garchy and Sens-Beaujeu (30 km west of Garchy).

The transmitted power is 4 kW. The transmitting antenna is a corner reflector, the two plane reflectors being respec-tively horizontal and vertical (Fig. 7.1). The beam is direct-ed eastward with an elevation of 45°; the 3 dB-beamwidth is $\pm$ 13° in azimuth and $\pm$ 10° in elevation; the antenna gain in the direction of the beam axis is 16 dB.

The three receiving antennae at Sens-Beaujeu are corner-reflectors, similar to the transmitting antenna, but of smaller size. Their beams are also directed eastwards with an elevation of 45°; their 3 dB beamwidth is $\pm$ 14° in azimuth and $\pm$ 20° in elevation; their gain is 14 dB. The receiver noise level is 10^{-13} milliwatts.

The zonal component of the wind is deduced from a Doppler shift measurement for each meteor echo. The device used for this purpose is illustrated in Fig. 7.1: the main transmitting and receiving antennae, T_1, R_1, R_2 and R_3 (mentioned above) are polarized horizontally and only detect the meteor echo. A vertically polarized communications link is used between T_2 and R_4 in order to have a wave transmitted directly from the transmitting to the receiving station.

To avoid a direct wave being received by R_1, R_2 and R_3 when high energy pulses are emitted by the main transmitter,

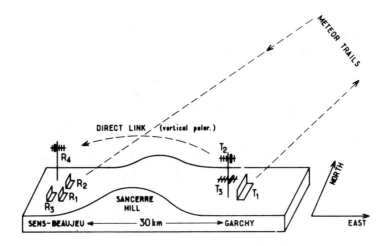

Fig. 7.1 The meteor radar at Garchy. T_1 : main transmitting antenna. R_1, R_2, R_3 : receiving antennae. T_2, R_4 : vertically polarized link for the transmission of the direct field. T_3 : transmitting antenna sending a wave with the same amplitude as T_1, but with opposite phase, in order to cancel the direct field received by R_1, R_2, R_3.

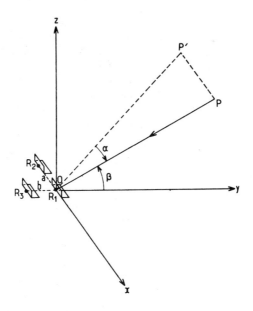

Fig. 7.2 Meteor radar measurement of the angles of arrival α and β of the received field.

the receiving and transmitting sites are located so that they
are hidden from one another by a ridge and the ground wave is
strongly attenuated. In addition, the backward gain of the
transmitting antenna is very small (-20 dB). Additionally, an
antenna, T_3, at the transmitting station sends a wave towards
the receiving station with the same amplitude as that of the
backward feedthrough, but with an opposite phase. Thus the
antennae R_1, R_2, R_3 receive only the meteor echoes, and R_4 receives only the direct field from the transmitter. The phase
difference ϕ between these two waves is then measured by a
phasemeter; the derivative $d\phi/dt$ yields the Doppler frequency
shift a much higher accuracy than that of the standard *beat
method*.

A complete localization of the reflection point on the
trail (i.e. the point where the wind is measured) is obtained
by determining (a) its direction and (b) its distance.

(a) the *direction of the echo* is obtained by measuring
the differences between the phases ϕ_1, ϕ_2, ϕ_3 received respectively by the antennae R_1, R_2, R_3 (Fig. 7.1). The relation

$$\phi\alpha = \phi_1 - \phi_2 = \frac{2\pi a}{\lambda} \sin \alpha$$

$$\phi\beta = \phi_1 - \phi_3 = \frac{2\pi b}{\lambda} \cos \beta \qquad (7.1)$$

where a is the distance R_1 R_2, b the distance R_1 R_3 (Fig. 7.2)
and λ the wavelength   yields the two angles α and β defining
the direction of the echo, α is the angle of this direction
with R_1 R_2 and $\pi/2 - \beta$ its angle with R_1 R_3. α and β closely
represent the azimuth and the elevation of the echo if α is
small. (We shall refer often to these angles as the
azimuth and the *elevation*, although this approximation is
never made. The exact definition of α and β is always used
in the calculations.)

The distances a and b between the antennae must be as
large as possible to improve the accuracy of the measurement,
but they must be small enough to prevent ϕ_α and ϕ_β having
variations greater than 2π when α and β vary within the beam
of the receiving antennae (if this were not so, the measurement
would be ambiguous). We have chosen a=20 m and b=30 m.

(b) the *distance of the echo* is obtained in principle by
transmitting simultaneously two waves with adjacent frequencies
f_1 and f_2:

$$E_1 \propto \cos (2\pi f_1 t + \phi_1)$$
$$E_2 \propto \cos (2\pi f_2 t + \phi_2)$$

Let $2r = r_1 + r_2$, r_1 being the distance from the reflection point on a meteor trail to the transmitting station, r_2 its distance to the receiving station (in the simple case when the distance between these two stations (30 km) is negligible, r represents the distance of the echo). The fields of these signals reflected by the trail and received by R_1 are

and

$$e_1 \propto \cos\ 2\pi f_1 (t-\frac{2\ r}{c}) + \phi_1'$$

$$e_2 \propto \cos\ 2\pi f_2 (t-\frac{2\ r}{c}) + \phi_2'$$

where c is the speed of propagation. The fields received directly in R_4, through the vertically polarized link, are

$$e_1' \propto \cos\ 2\pi f_1 (t-\frac{2\ r}{c}) + \phi_1''$$

$$e_2' \propto \cos\ 2\pi f_2 (t-\frac{2\ r}{c}) + \phi_2''$$

where $2\ r_o$ is the distance between the transmitting and the receiving stations. After mixing e_1' and e_2', we can filter out a component proportional to

$$\cos\ \{2\pi (f_2-f_1)(t-\frac{2\ r}{c}) + \phi_2''-\phi_1''\ \}$$

which is then mixed with e_1, thus giving a wave e of frequency f_2. The phase difference between e and e_2

$$\phi_D=\frac{4}{c}\pi(f_2-f_1)(r-r_o) + \phi_2''-\phi_1''-\phi_1'-\phi_2' \tag{7.2}$$

can be measured and yields the distance r.

The difference f_2-f_1 must be as large as possible to improve the accuracy of the measurement, but small enough to prevent ϕ_D to have variations greater than 2π when r describes its range. The value $f_2-f_1 = 600$ Hz thus chosen yields r with a poor accuracy (see Section 7.3). To improve this, a third frequency f_3 was used, such that the difference $f_1 - f_3$ is much larger than $f_2 - f_1$. Operating on f_1 and f_3 on the same way as on f_2 and f_1 one gets a phase difference,

$$\phi_{DV} = \frac{4\ \pi}{c}\ (f_1 - f_3)(r - r_o) + \phi_1''-\phi_3'' + \phi_1' - \phi_3' \tag{7.3}$$

which gives a more precise, but ambiguous, determination of r. However, the ambiguity can be resolved, since ϕ_D is known: the measurement of ϕ_{DV} plays the role of a *vernier*. The frequencies chosen are f_1=29796.8, f_2=29797.6, f_3=29792.0 kHz.

The Doppler shift $d\phi/dt$, the direction of the echo and its distance are then deduced from phase measurements. As a first

step (during the period 1965-67), all these phase comparisons
were made by optical *phase comparators* which produced a phase-
amplitude diagram on the screen of an oscilloscope (Delcourt
et al., 1964; Spizzichino et al., 1965; Revah, 1968). Photographic records were then used, but the need of a visual processing method seriously limited the use of the equipment.
However, 10 recording runs of 48 to 72 hrs (1965-66) were so
analyzed and yielded the preliminary results described below
(Section 7.6).

Since 1969, phasemeters have been used, and their output
voltages are recorded in analog fashion on magnetic tapes. Two
examples of such records are given in Fig. 7.3. These analog
records are then converted into digital records and directly
analyzed by a computer. A program has been worked out in
order to recognize the meteor echoes and extract the Doppler
shift, azimuth, elevation and distance. It yields an estimation of the error for each of these parameters and eliminates
the echo whenever one of these errors is too large.

7.3 Calibration and Errors

The angles α and β are deduced from the phase differences
ϕ_α and ϕ_β given by (7.1) but two corrections must, in fact,
be made to the theoretical formulae (7.1):

(a) the phase differences ϕ_α and ϕ_β are obviously not
measured at the output of the antennae but at the output of
the receivers (the phase comparison is made between narrow-
bandwidth intermediate frequency of the receivers, all of them
sharing the same local oscillator): we must then take account
of the phase variations Ψ_α and Ψ_β produced respectively by the
equipment (circuits, connections, etc...).

(b) The theoretical formulae (7.1) do not take account of
the mutual influence between the three antennae R_1, R_2, R_3.
Small corrections $\Delta\alpha$ and $\Delta\beta$ must then be applied to the values
of α and β deduced from 7.1, which may be written

$$\phi_\alpha = \frac{2\pi a}{\lambda} \sin\{\alpha - \Delta\alpha(\alpha,\beta)\} + \Psi_\alpha(t)$$

$$\phi_\beta = \frac{2\pi b}{\lambda} \cos\{\beta - \Delta\beta(\alpha,\beta)\} + \Psi_\beta(t) \qquad (7.4)$$

Ψ_α and Ψ_β are slowly varying with time; they depend on the
stability of the electronic components, the temperature variations, etc. They can be measured by using a fixed source (a
small dipole) situated behind the antennae; the phase of the
echoes is then compared with this fixed phase-reference. In

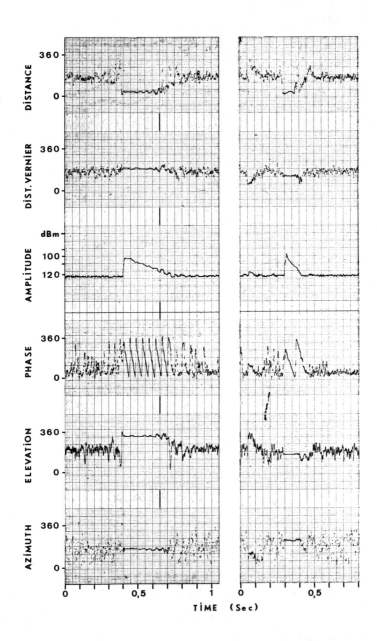

Fig. 7.3   Two examples of meteor echo records obtained at
Garchy on Jan. 31, 1970.

practice, the latter was measured once an hour during each
recording run; furthermore, four different sources were always
used successively for the sake of verification. As shown on
Fig. 7.4, the variations of the four phase-references do not
generally exceed a few degrees per hour and are almost parallel.   It must be mentioned that many precautions were taken
to obtain this result (climatization; elimination of mutual
influences between the electronic circuits, etc.)

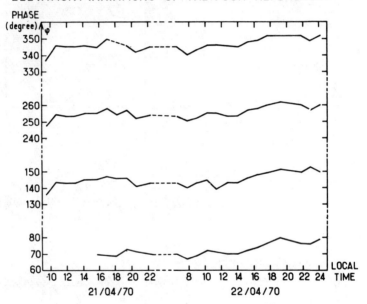

Fig. 7.4 Example of elevation measurement variations of the
four phase references.

  Δα and Δβ do not vary with time; they only depend on the
geometry of the antennae and on the angles of arrival α and
β of the received field. These corrections can be determined
empirically by using a source moving within the beam of the
receiving antennae. This calibration has been made by means
of a small transmitter hung below a helicopter and is described in Appendix III of (Revah, 1968). Some care must be
taken to obtain consistent results: the distance between the

helicopter and the source must be large enough (about 150 m)
to prevent its radiation being perturbed by the metallic body
of the helicopter; the source must remain far enough (about
1 km) from the antennae to avoid proximity effects. The following results were obtained:

(a) The corrections Δ_α and Δ_β remain almost constant
during a time interval of 2 years; their variations between
two successive calibrations in 1965 and 1967 were of the order
of 0.25°.

(b) Δ_α can be neglected with a very good approximation Δ_β
can be represented by a linear function of α:

$$\Delta\beta = \Delta\beta_0\ (\beta) + p\ (\beta)\alpha$$

within an error interval of 0.25°. The variations of the axial
correction $\Delta\beta_0$ and of the slope p and β are shown in Fig. 7.5.
The correction $\Delta\beta$ never exceeds 1.5°.

The *noise* superposed on the meteor echoe signals is one
of the main causes of error. Let us consider the general case
of two fields S_1 and S_2 of respective phases ϕ_1 and ϕ_2, superposed respectively on noises N_1 and N_2. It can easily be
shown that the phases of the resultant signals fluctuate with
respective standard deviations:

$$\delta\phi_1 = \frac{N_1}{S_1} \qquad \delta\phi_2 = \frac{N_2}{S_2}$$

and that the phase difference $\phi_1 - \phi_2$ measured by a phasemeter
or an optical *phase comparator* fluctuates with a standard
deviation

$$\delta\phi = \{\delta\phi_1^2 + \delta\phi_2^2\}^{\frac{1}{2}} = \{(\frac{N_1}{S_1})^2 + (\frac{N_2}{S_2})^2\}^{\frac{1}{2}} \qquad (7.5)$$

(Revah, 1968). This result can be applied to the determination
of the angles of arrival: the phase differences ϕ_α and ϕ_β
between the same meteor echo received by two different antennae
fluctuate with standard deviations,

$$\delta\phi_\alpha = \delta\phi_\beta = \sqrt{2}\ \frac{N}{S} \qquad (7.6)$$

For a signal/noise ratio S/N of 25 dB, we find $\delta\phi_\alpha = \delta\phi_\beta = 4.5°$.
We shall consider this value as an upper limit of the error
over ϕ_α and ϕ_β, although in practice this error is generally
much smaller: most of the meteor echoes have a duration that

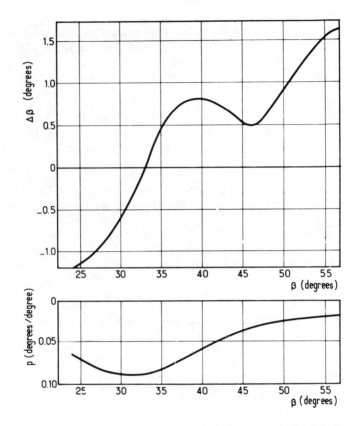

Fig. 7.5 Correction $\Delta\beta_o$ and coefficient p deduced from the calibration of the geometry of the system, make by means of a small transmitter hung below an helicopter.

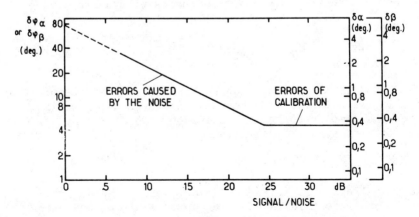

Fig. 7.6 Error of the angles of arrival.

is long compared to the characteristic time of the fluctuations of ϕ_α and ϕ_β so that their mean values ϕ_α and ϕ_β can be determined with an error that is much smaller than the standard deviations $\delta\phi_\alpha$ and $\delta\phi_\beta$.

The errors $\delta\alpha$ and $\delta\beta$ corresponding to $\delta\phi_\alpha = \delta\phi_\beta = 4.5°$ are given by (7.1). For $\alpha = 0$ and $\beta = 45°$ we obtain

$$\delta\alpha = 0.35 \qquad\qquad \delta\beta = 0.3 \qquad\qquad (7.7)$$

As noted above, the phase deviations Ψ_α, Ψ_β produced by the equipment never exceed 5° between two successive calibrations (i.e. in one hour); they are then of the same order as the errors $\delta\phi_\alpha$, $\delta\phi_\beta$ caused by the noise for a signal/noise ratio of 25 dB. We have seen that the systematic corrections $\Delta\alpha$, $\Delta\beta$ are determined within an accuracy of 0.25°; they are then slightly smaller than the errors $\delta\alpha$, $\delta\beta$ caused by the noise for a signal/noise ratio of 25 dB.

Other sources of error have been analyzed in detail in (Revah et al. 1962; Revah, 1968) and are considered to be negligible. A small shift of the transmitted frequency would give rise to a smaller phase shift in the lowest intermediate frequency of the receivers; as the oscillators used at the transmitting and receiving ends have a relative stability of 10^{-7}, the resulting errors of ϕ_α and ϕ_β do not exceed 1°.

Another source of error is the atmospheric refraction which can bring about a slight difference between the true direction of the trail and that of the received field. The error caused by tropospheric refraction, calculated after (Schulkin, 1952) and (Bean, 1960), never exceeds 0.05°; that caused by ionospheric refraction in the D and E regions are found to be of the same order of magnitude.

In conclusion, the error over α and β is proportional to the signal/noise ratio if it is smaller than 25 dB, and remains of the order of 0.3° for echoes reaching more than 25 dB above the noise (Fig. 7.6).

The distance r is deduced from the phase ϕ_D and the _vernier_ ϕ_{DV} by (7.2) and (7.3), where each of the differences $\phi_1' - \phi_2'$, $\phi_1'' - \phi_2''$, $\phi_1' - \phi_3'$, ϕ_1'' ϕ_3'' is the sum of two terms.

(a) The difference between the phase shifts caused by the transmitting and receiving antennae and by the connections between these antennae and the transmitter and the receiver. None of these circuits is selective enough to generate different phase shifts for the frequencies f_1, f_2 and f_3; therefore,

corresponding terms of the phase difference can then be neglected.

(b) The difference between the phase shifts caused by the
receiving circuits. In order to determine these terms, the
echo receiver and the *direct wave* receiver (which are normally
connected to the antennae R_1 and R_4, respectively) are fed in
parallel by the direct wave antenna R_4. The obtained values
ϕ_D and ϕ_{DV} of ϕ_D and ϕ_{DV} correspond to $r = r_o$, so that (7.2)
and (7.3) may be written

$$\phi_D = \frac{4\,\pi}{c}(f_2 - f_1)\ (r - r_o) + \phi_{D_o}$$

$$\phi_{DV} = \frac{4\,\pi}{c}(f_1 - f_3)\ (r - r_o) + \phi_{DV_o}$$

This calibration was made once per hour during each recording
run; ϕ_D and ϕ_{DV} were found to vary by less than $3°$ between
two successive calibrations.

The range *error caused by noise* can be calculated in the
same way as for the angle of arrival (Revah, 1968). The
standard deviations $\delta\phi_D$ and $\delta\phi_{DV}$ are given by (7.5), where
N_1/S_1 is the ratio N/S of the echo, and N_2/S_2 the signal/noise
ratio of the direct wave, which is very small. Hence

$$\delta\phi_D = \delta\phi_{DV} \simeq \frac{N}{S} \qquad\qquad (7.8)$$

For a signal/noise ratio of 25dB, we find $\delta\phi_D = \delta\phi_{DV} = 3°$. For
the same reason mentioned above relating to the angle of arri-
val, this value represents an upper limit of the error caused
by the noise.

The error δr, corresponding to $\delta\phi_D$ for a signal/noise ratio
of 25 dB, given by (7.2), is 2 km, but the *vernier* would yield
r, from (7.3), with an accuracy

$$\delta r = 0.3 \text{ km} \qquad\qquad (7.9)$$

The *errors caused by calibration* depend only on the variations of ϕ_{D_o} and ϕ_{DV_o} between two successive calibrations. We
have seen above that it never exceeds $3°$. The corresponding
error over r is $\delta r = 0.25$ km.

Other sources of error have been analyzed in detail in
(Revah et al. 1962; Revah 1968). They are similar to those
concerning the angles of arrival and are also considered to be
negligible. The phase shift caused by the small frequency
shifts in the receivers can be reduced to $1°$ or less, provided that the transmitting and receiving oscillators have a

stability of 10^{-4}. Tropospheric refractions lead to errors smaller than 10 m. The errors caused by ionospheric refractions in the E-layer can reach 400 m at the upper edge of the meteor range but are, in general, much smaller.

In conclusion, the error in r is proportional to the signal/noise ratio if it is smaller than 25 dB and remains of the order of 0.3 km for echoes reaching more than 25 dB above the noise (Fig. 7.7).

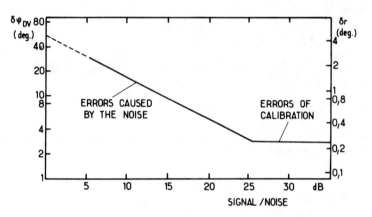

Fig. 7.7   Error in range measurement.

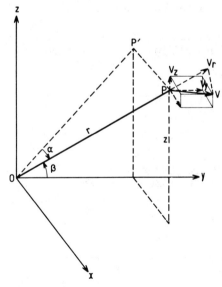

Fig. 7.8 Determination of the altitude in the simple case of a plane earth, the distance between the transmitting and the receiving station being neglected.

The correct way of deducing the altitude z from the measured values of α, β, and r is described (Section 7.5), but a rough estimation of the error in z can be obtained easily by neglecting the earth curvature and the distance r_0 between the transmitting and the receiving ends. As can be seen from Fig. 7.8, z is given by

$$z = r \cos \alpha \sin \beta \qquad (7.10)$$

and the error δz by

$$\delta z = \delta r \sin \beta \cos \alpha + z \cot g \, \beta \delta \beta + z |tg \, \alpha| \delta \alpha. \qquad (7.11)$$

z varies within the 80-105 km range; α and β are limited by the transmitting antenna beam (see Section 7.2)

$$|\alpha| < 13° \qquad\qquad 35° < \beta < 55°.$$

The maximum value of δz can be deduced from these limits. For a signal/noise ratio of 25 dB, $\delta \alpha$, $\delta \beta$, and δr are given by (6) and (8). We find

$$\delta z \quad 0.9 \text{ km}$$

Starting from the variations of $\delta \alpha$, $\delta \beta$, and δr given in Figs. 7.6 and 7.7, the variations of δz with respect to the signal/noise ratio can be deduced and are given in Fig. 7.9.

As mentioned above, $\delta \alpha$, $\delta \beta$, and δr are overestimated values of the errors in α, β and r; δz is then an overestimated value of the error in z.

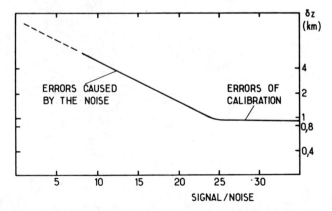

Fig. 7.9 Error in altitude (z), deduced from the errors in α, β and r.

Any error smaller than 1 km in the altitude has no physical interest. We must keep in mind that the reflected field
does not come from one point but from the whole first Fresnel
zone which extends within an altitude range of the order of 1
km. Details of the wind structure smaller than 1 km are always
smoothed out in any radio meteor measurement.

The correct way of deducing the *zonal* wind velocity from
the phase rotation $d\phi/dt$ of the meteor echoes is described
below (Section 7.5); we shall obtain here a rough estimation
of the error (as above) by neglecting the earth curvature and
the distance r_0 between the transmitting and the receiving
ends. The radial velocity u_r is then given by

$$u_r = \frac{\lambda}{4\pi} \frac{d\phi}{dt} \qquad (7.12)$$

u_r depends generally on the three coordinates u_x, u_y, u_z of
the wind velocity. With the y-axis directed eastwards and the
z-axis upwards (Fig. 7.8),

$$u_r = u_x \sin \alpha + u_y \cos \alpha \cos \beta + u_z \cos \alpha \sin \beta \qquad (7.13)$$

The antenna beams being directed eastwards (α small), and the
vertical wind velocity u_z being small, we shall assume as a
first approximation that u_r only depends on the zonal wind u_y

$$u_r \simeq u_y \cos \alpha \cos \beta \qquad (7.14)$$

The value of $d\phi/dt$ is deduced from 10 successive values of ϕ,
at intervals of 10^{-2} sec, each of them having from (7.5) an
error $\delta\phi = N/S$. The slope $d\phi/dt$ of the regression line is then
determined within on error of 0.5 sec^{-1} for a signal/noise
ratio of 25 dB. The error of the radial velocity u_r given
by (7.12) is then

$$\delta u_r = 0.5 \text{ m sec}^{-1}$$

The main cause of error in determining the *zonal* velocity
is due to the terms proportional to u_x and u_z neglected in
(7.14). Evaluations of the vertical wind u_z have been obtained
by chemiluminescent trails (Edwards et al. 1963) and from
meteor observation (Roper, Elford, 1965): it is generally
of the order of 5 m sec^{-1} at meteor heights. The corresponding
error over u_y is of the same order.

The error resulting from the term $u_x \sin \alpha$ is small if
α is small. With a transmitting antenna beam-width of $13°$
in azimuth this term reaches 0.22 u_x at the edge of the beam:
with u_x = 50 m sec^{-1} it would reach 11 m sec^{-1}. Such an
error is unacceptable, but we shall see in Section 5 that

a. the error can be reduced by eliminating all the echoes with too large values of $|\alpha|$, without losing too large a proportion of the obtained date;

b. the above value represents the error over the wind given by each *individual* meteor; the error over the interpolated wind can be much smaller, provided that the hourly rate of the data is high enough. (It must also be noticed that this error results from the approximation (7.14) which is only needed if a single radar is available to measure the zonal component of the wind. This is the case for the preliminary results obtained in Garchy and described in Section 6, but a second radar will be operating from the end of 1970 to measure the meriodional component of the wind in the same region. The data of each of these radars will enable us to correct the wind measurement made by the other and therefore make it possible to produce wind measurements within an accuracy of a few m sec^{-1}.)

7.4 Hourly Rate of the Obtained Data-Comparison with Other Experiments

We have seen in Section 7.3 that a wind measurement is yielded at a point localized within $\pm$ 1 km, by each meteor echo,the signal/noise ratio of which is larger than 25 dB. The noise level being -130 dBm under normal conditions, this accuracy is obtained for echoes reaching -105 dBm.

The mean hourly rate of echoes reaching this level is given in Fig. 7.10. The dashed line on the same figure gives

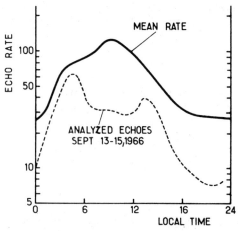

Fig. 7.10 Mean hourly rate of echoes of level larger than -105 dBm (full line); hourly rate of analyzed echoes on Sept. 13-15, 1966 (dashed line).

the effective mean rate of analyzed echoes for a particular
recording run (September, 1966) which represent only 30% to
70% of the total number of echoes. This loss of echoes results from many different causes, and we can outline only briefly the most important ones. (More details are given in
(Revah et al., 1966) for the photographic recording system
used in the first recording runs.)

(a) A large proportion of the received echoes (30-60%)
were lost in 1965-1966 because of the photographic recording
method used then. Successive events were sometimes superposed on the phase-amplitude diagrams, or a meteor echo arrived
between two successive photographs, etc.

(b) Airplane echoes can easily be distinguished from
meteor echoes, but they blind out the radar about 20% of the
time. (A special device has been made to eliminate them (as
for all echoes from near targets) (Glass et al., 1969), and
should be operating from the end of 1970).

(c) Backscatter echoes from the F-region were observed in
the daytime during the years of strongest solar activity (1967-
70). They can reach -115 dBm between 9 and 12 a.m. In this
extreme case the number of usable meteor echoes is reduced by
a factor of about 20. The best protection against the F-region
echoes is obtained by using a more directional transmitting
antenna. (A first attempt made in 1969 showed that these F-
region echoes could be reduced very easily by 5 dB; but this
study has been provisionally stopped (until 1978) owing to the
decrease of solar activity.)

The difficulties described in (b) and (c) can be summarized by stating that the Garchy radar has the main disadvantages of any C. W. system - i.e., the lack of protection
against interference and the impossibility of separating two
simultaneous but different echoes. We have seen that these
disadvantages reduce the echo rate, but generally not in a
crucial way. The meteor echoes having a short duration are
received successively and do not need to be separated from
each other.

On the other hand, Sections 7.3 and 7.4 show that the
equipment at Garchy yields *both* a high hourly rate of data
and a good sensitivity: these two advantages were never simultaneously present in previous experiments using pulsed radars
(for instance, Jodrell Bank, Greenhow and Neufeld, 1956, 1961).
One may then wonder whether the use of continuous waves provides another decisive advantage.

Pulses need wide-band receivers, which have then a poor sensitivity. This loss of sensitivity is theoretically compensated by the possibility of having a higher transmitted power during the short time interval of the pulse: it can be shown that a pulse and a CW radar can reach the same sensitivity if they have the same mean transmitted power. It is well known that it is far easier (or less expensive) to build a 5 kW CW transmitter than a pulsed one giving a mean power of 5 kW (and then a peak power of a few megawatts). For the same price (or work), a CW radar may then reach a higher sensitivity than a pulsed one.

For example, the pulse radar of Jodrell Bank (Greenhow and Neufeld, 1956, 1961) used pulses with a peak power of 50 kW, and a receiver band-width of 100 kHz. The ratio of the transmitted power to the sensitivity of the receiver is then about 10^{18}. This ratio is $1.5 \ 10^{19}$ for the Garchy CW radar.

One can make use of such an advantage in two different ways: either the higher sensitivity of the equipment is used to obtain a higher rate of received echoes, or an equivalent number of echoes is analyzed but with a larger signal/noise ratio, and then a higher accuracy in the location and wind speed measurement. We have chosen this second possibility in Garchy (although our radar makes it possible to choose the first one, or any intermediate solution). (When keeping only the echoes of signal/noise ratio larger than 25 dB, a height definition of 0.8 km is obtained (Fig. 7.8), with an echo rate given by Fig. 7.9. But that system has in fact a great flexibility: for instance, the echoes of S/N ratio higher than 15 dB would yield a height definition of 2.5 km and echo rates about 6 times larger than those given in Fig. 7.9.)

Before concluding this comparison with other experiments, let us mention that our measurement of the phase rotation $d\phi/dt$ (Section 7.2) appeared much more suitable than frequency shift measurements for meteor wind observations. In many experiments (Jodrell Bank for instance), the frequency shift is deduced from the beat frequency of the echo interfering with the direct wave. This is only possible for echoes of duration T at least of the same order as the beat period given after (7.12) and (7.14) by

$$\tau = \frac{\lambda}{2 \ u_y \cos \alpha \cos \beta}$$

The wind measurement is only possible if

$$u_y > \frac{\lambda}{2T \cos \alpha \cos \beta}$$

With T = 0.5 sec, $\alpha=0$, $\beta=45°$, $\lambda=10$ m, we find

$$u_y > 15 \text{ m sec}^{-1}$$

Wind values smaller than about 15 m sec^{-1} are fairly often eliminated by frequency shift measurements; the wind distribution thus obtained is then cut down. On the contrary, the phase rotation method we have used makes it possible to measure winds of the order of 1 m sec^{-1} (Section 7.3).

7.5 Data Processing

The first three parts of this section deal with three successive steps of the data reduction. We shall describe the reduction of the data for each individual echo, i.e., the process used to deduce the zonal wind u_y and the coordinates x, y, z of the reflecting point from the primary data $\phi\alpha$, ϕ_β, ϕ_D, ϕ_{DV}, $d\phi/dt$. We shall discuss the interpolation method to be used to deduce the variations of the wind profile from these individual data and the methods used to analyze these profiles. The harmonic analysis appears more convenient, and enables us to separate different components of the wind (prevailing wind, tides, gravity waves). The last paragraph describes an empirical verification of the methods of analysis given above, using synthetized wind profiles.

The first problem which arises is to choose the appropriate time interval to pick up the primary data ϕ_α, ϕ_β, ϕ_D, ϕ_{DV}, $d\phi/dt$. All of them vary slightly during the lifetime of a meteor echo; during the first hundredth of a second of this lifetime, the trail formation is not completely achieved, and any measurements would be perturbed by a diffraction effect (MacKinley, 1961, Chap. 8). About 0.3 sec after its formation, the trail becomes distorted and can present more than one reflection point (Greenhow, 1950 and 1952), (Manning 1959), (Kent, 1960), (Revah, Spizzichino, 1963 and 1964). The examples given in Fig. 8.11 show indeed that the value of the slope $d\phi/dt$ is more or less perturbed at the beginning and at the end of the echoes. A systematic study of these perturbations leads us to pick up only the primary data in the time interval ranging between 0.05 and 0.20 sec after the echo rise, where they remain generally constant.

Let us now compute the three coordinates x, y, z of the reflecting point P, with the earth tangent plane as a first approximation (Fig. 7.12). The receiving station R is taken as the axis origin; as in Section 7.3 above, the y-axis is directed eastwards, and the z-axis upwards. The transmitting station T is practically on the y-axis, with an ordinate

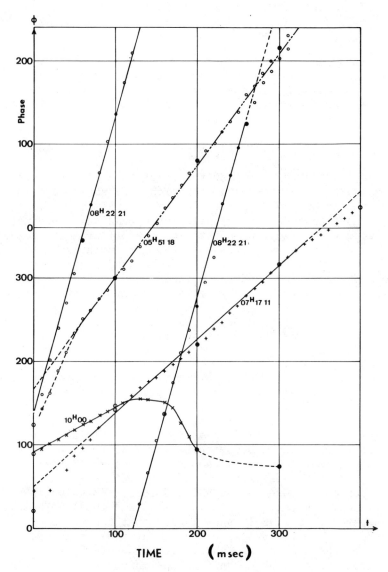

Fig. 7.11 Examples of phase rotation of the detected echoes
(Garchy, 9 Sept. 1965). The phase varies generally linearly
with height, but its variations are always perturbed at the beginning (diffraction effect) and at the end (distorsion effect)
of the echo. In one particular case (10^H 00), the distorsion
effect appears earlier, but such a case is rather scarce. In
order to obtain a correct wind measurement, the phase variation
is picked up only in the time interval ranging between 0.05
and 0.20 sec after the echo rise.

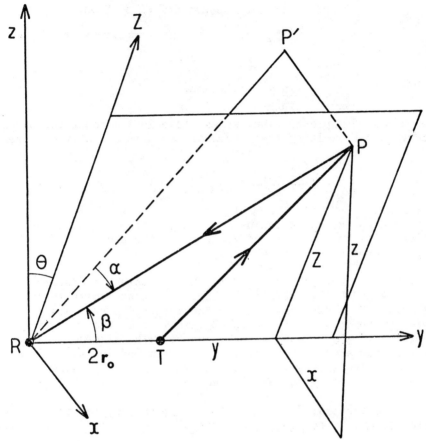

Fig. 7.12 Determination of the altitude in the case of a plane earth.

$RT = 2 r_o = 30$ km. As we take account of the separation between the transmitting and the receiving stations, the measured distance r represents one half of the echo path

$$r = 1/2 \ (EP + PR)$$

The coordinates y and Z of P in a plane containing Ry and P are deduced from r and β by the following set of equations:

$$\frac{(y - a)^2}{r^2} + \frac{Z^2}{r^2 - r^2} - 1 = 0$$

$$y \ tg \ \beta - Z = 0$$

x and z are then deducted from α, β and Z by

$$x = Z \sin \theta \qquad\qquad y = Z \cos \theta$$

with

$$\sin \theta = \frac{\sin \alpha}{\sin \beta}$$

As a second approximation, we can take into account the earth curvature by writing that the altitude h is slightly different from z:

$$h = z + \frac{x^2 + y^2}{2R} = z + \frac{y^2}{2R}$$

where R is the earth radius. The correction $y^2/2R$, in practice, lies between 0.5 and 2.5 km, and the neglected fourth-order terms never exceed 0.1 km.

Figures 7.13 and 7.14 show examples of distributions of x, y and h thus obtained. Their horizontal distribution (Fig. 7.13) shows that most of the detected echoes are located within the section of the 10 dB main lobe of the transmitting antenna by a sphere of altitude 95 km. The sizes of this region are about 140 km along the y-axis and 50 km along the x-axis. Their vertical distribution (Fig. 7.14) shows that they range mostly between 80 and 110 km.

The zonal wind u_y is deduced from the Doppler phase rotation $d\phi/dt$. by neglecting the vertical component u_z. We have then

$$\frac{d\phi}{dt} = \frac{4\pi}{\lambda} \left(\frac{x}{r} u_x + \frac{y - r_o}{r} u_y\right)$$

For small values of x, the term $\frac{x}{r} u_x$ can be neglected, with an error over u_y:

$$\delta u_y = \frac{x}{y - r_o} u_x$$

The preliminary results described in Sec. 7.6 have been obtained by leaving out all the echoes having a ratio $|x|/(y-r_o)$ 0.25. The distribution of $x/(y-r_o)$ given on Fig. 7.15 shows that 16% of the echoes are so lost. The r.m.s. value of u_x being of the order of 30 ms^{-1} at 90 km and 50 ms^{-1} at 100 km, we have

$$|\delta u_y| < 7.5 \text{ m sec}^{-1} \text{ at 90 km} \qquad (7.15)$$

$$|\delta u_y| < 12.5 \text{ m sec}^{-1} \text{ at 100 km}$$

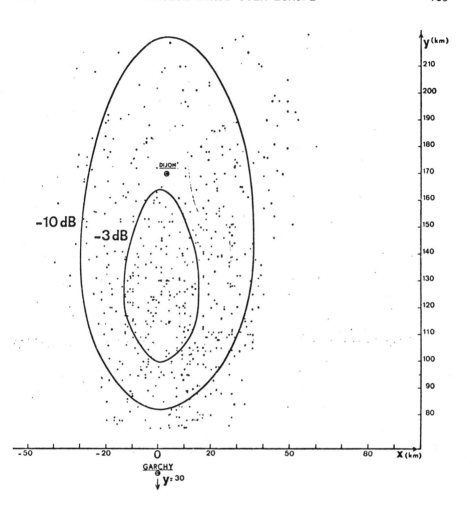

Fig. 7.13 Example of horizontal distribution of the received echoes (Dec. 1965)

$|\delta u_y|$ is the r.m.s. error over each individual wind measurement. As shown below, the error over the interpolated wind profile can be smaller than $|\delta u_y|$, if the data are sufficiently superabundant.

The second step of the data processing consists in using the wind values provided by the individual echoes to reconstitute the wind variations in time and space.

It will be assumed, as a first approximation, that the wind does not depend on the horizontal coordinates x and y. This hypothesis has already been introduced by all the authors

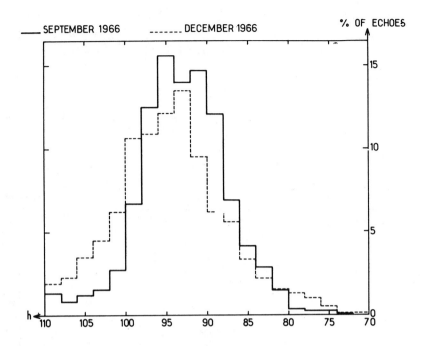

Fig. 7.14 Examples of vertical distribution of the received echoes (Dec. 1965, Sept. 1966).

who have described the analysis of meteor wind observations. Furthermore, it will be verified a posteriori (Section 7.6) for most of the wind oscillations observed by the radar of Garchy; they are found to have horizontal wavelengths that are much larger than the dimensions of the observed region (shown in Fig. 7.13). (Nevertheless, this hypothesis does not represent a basic limitation of the observing system, which yields the horizontal coordinates of the echoes and then could be used to obtain rough estimations of the horizontal wind gradient.)

We have then a series of individual values of time, height and zonal wind yielded by each echo t_i, h_i and u_i, and wish to obtain the best estimation of the continuous function $u(t,h)$ for any particular value t_0, h_0 of t and h.

The solutions of such a problem are well known in the simple case of a single variable function $u(t)$. The simplest one consists in a linear interpolation: $u(t)$ is supposed to be linear between two consecutive points t_i and t_{i+1}, with

$$t_i < t_0 < t_{i+1}.$$

DISTRIBUTION OF $x/y\text{-}r_0$

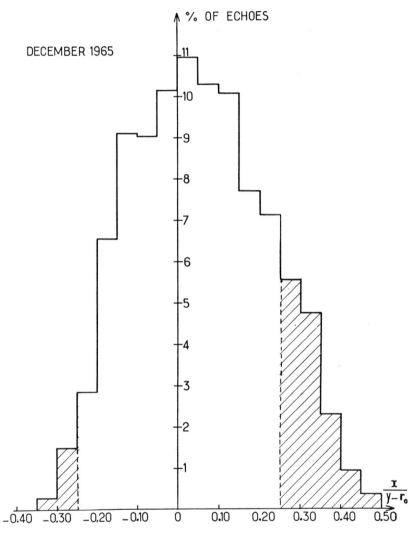

Fig. 7.15　Example of distribution of x/y-r₀.

In general this will be an over simplification, and it will be necessary to consider this matter in more detail. Of special importance are the population density of the data and the errors associated with those data. It is necessary, then, to devise analysis techniques which will provide representative results in cases where the data are adequate as well as when the data are inadequate.

Two difficulties arise for our present application:

(a) u is a function of two variables. There is no basic objection to the linearization of U (t, h) in a small region around any given point t_o, h_o where we want to obtain an estimation of u. The variations of u (t, h) are then represented by a plane which must be defined by 3 points (i.e., 3 meteor echoes). But how is one to choose these three points? The method for a one-variable function, which consisted in selecting two consecutive points around t_o, is no longer available; the word consecutive has no meaning in a two-dimensional space. One could intuitively suggest the choice of the three points which are at the smallest distance from the point t_o, h_o; but t is a time and h a length. A distance in the (t, h) plane depends on the chosen units and has no physical meaning. This difficulty is smoothed out in the method described below by using the two-dimensional autocorrelation function $\rho(t-t_o, h-h_o)$ of the zonal wind instead of the distance: we could choose the three points t_i, h_i for which the autocorrelation function $\rho(t_i-t_o, h_i-h_o)$ is the greatest.

(b) Each u is determined with an error of the order of 10 m/s; but as we often have a superabundant density of data, it is then obvious that a least-mean-square method, using more than 3 experimental points to determine the linearized function u(t, h), is more satisfactory.

First, let us for the sake of simplicity, describe this method for a one-variable function and look for an interval $(t_o-\Delta t, t_o + \Delta t)$ (if it exists) for any particular value t_o of the time where

$$u(t) = u(t_o) + u'(t_o) (t-t_o)$$

$u(t_o)$ can be determined by the standard least-mean-square method of the regression line, provided that at least two experimental data exist within this interval.

This approximation is only valid if the second-order term can be neglected:

$$\left| 1/2 \ u''(t_o)(t-t_o)^2 \right| << \left| u'(t_o)(t-t_o) \right|$$

which is equivalent, to the third order, to

$$1/2 \left| u'(t)-u'(t_o) \right| << \left| u'(t_o) \right| \qquad (7.16)$$

Let $\tau = t-t_0$. It is generally not possible to find a value of Δt such that (7.16) is satisfied for any t_0, by any t ($t_0-\Delta t$, $t_0+\Delta t$), i.e., for any τ ($-\Delta t$,$+ \Delta t$). But (7.16) can be satisfied for most t_0 if

$$1/2(u'(t_0 + \tau) -u'(t_0))^2 << u'(t_0)^2$$

for any τ ($-\Delta t$, $+ \Delta t$) (here, $u'(t)$ is supposed to be stationary so that the averages do not depend on t). This inequality may be written

$$1 - \rho(\tau) << 2 \qquad \text{if } |\tau| < \Delta t \qquad (7.17)$$

where $\rho_t(\tau)$ is the autocorrelation function of the derivative $u'(t)$; (7.17) means that Δt must be small compared to the autocorrelation radius of $u'(t)$.

We must than know the autocorrelation function $\rho_t(\tau)$ before interpolating. It can be shown that $\rho_t(\tau)$ is proportional to the second derivative of the autocorrelation function $\rho(\tau)$ of the wind $u(t)$ (see, for instance, J. Stern et al., Methodes pratiques d'etudes des fonctions aleatoires, Dunod, Paris, 1967):

$$\rho(\tau) = - \frac{\rho''(\tau)}{\rho''(0)}$$

$\rho(\tau)$ can be directly estimated from the series of experimental data t_i, u_i:

$$\rho(\tau) \simeq \frac{1}{\sigma^2} \dot{\Sigma}_i (u_i-\bar{u})(u_j-\bar{u})$$

the sum being extended to all the couples u_i, u_j corresponding to $t_i-t_j = \tau$, and the average u and the standard deviation σ of $u(t)$ being also deducted from the series of experimental data.

One could believe that the problem is now solved: two successive derivations of $\rho(\tau)$ deduced from experimental data yield $\rho_t(\tau)$, and then its autocorrelation radius Δt. In practice, the successive derivations would lead to a drastic loss of accuracy. But if we try to represent $\rho(\tau)$ by a mathematical model, we notice the most convenient one is always intermediate between a sinusoidal function (wave motion) and a Gaussian function (random motion). The condition (7.17) written in the form

$$1 - \rho_t(\tau) < \varepsilon \qquad (7.18)$$

(ε small) leads in both cases to the condition:

$$1 - \rho(\tau) < \varepsilon' \qquad\qquad (7.19)$$

For a sinusoidal model, $\rho_t(\tau) = \rho(\tau)$, so that $\varepsilon' = \varepsilon$. For a Gaussian model (and other similar ones, like $\rho = 1/(1 + \tau^2)$, etc) ε' is always smaller than ε, but of the same order of magnitude ($\varepsilon' = \frac{1}{3}\varepsilon$ for the Gaussian law).

We are then led to use a practical interpolation method in two successive steps:

    (a) the deduction of the autocorrelation function $\rho(\tau)$ from the experimental data,

    (b) the use of (7.19) instead of (7.18) to determine the interpolation interval $(t_o - \Delta t, t_o + \Delta t)$, with a value of ε' between $\frac{\varepsilon}{3}$ (Gaussian law) and ε (sinusoidal law). (The exact value of the ratio ε'/ε is not critical; the error is proportional to $\sqrt{\varepsilon}$, as shown below).

This method can be easily extended to our practical case where the wind u(t, h) is a function of two variables. The condition for the linear approximation being valid is obtained by the same process as above and leads to two conditions similar to (7.17):

$$1 - \rho_t(\tau, \zeta) \ll 2$$

$$1 - \rho_h(\tau, \zeta) \ll 2$$

where ρ_t and ρ_h are respectively the autocorrelation functions of $\partial u/\partial t$ and $\partial u/\partial h$ which are found to be proportional to the second derivatives $\partial^2 \rho/\partial \tau^2$ and $\partial^2 \rho/\partial \zeta^2$ of the two-dimensional autocorrelation function of the wind $\rho(\tau, \zeta)$. The two conditions similar to (7.18)

$$1 - \rho_t(\tau, \zeta) < \varepsilon$$

$$1 - \rho_h(\tau, \zeta) < \varepsilon$$

are found to be equivalent to the condition (7.19).

    The same steps (a) and (b) mentioned above can then be used for a practical calculation: the first one yields $\rho(\tau, \zeta)$ from the experimental data; the second yields an interpolation region around the point t_o, h_o in the (t, h) plane (instead of an interpolation interval). The interpolation is only possible if at least three experimental data exist in this region. A regression plane can then be determined.

    Errors. One can obtain an evaluation of the error caused

by linearization, if the condition (7.18) is satisfied for a
given ε. An order of magnitude Δu of the neglected second-
order terms is found by following the same process as above to
obtain equations (7.16) and (7.17), (The error is computed here
for a one-variable function u (t). The same process, and the
results, remain valid for a function of two variables.)

$$\Delta u^2 \sim 1/2\{1-\rho_t(t-t_o)\}\overline{u'(t_o)^2}\ (t-t_o)^2$$

$$< \sqrt{\frac{\varepsilon}{2}}\ \overline{u'(t_o)^2}\ (t-t_o)^2$$

Here $\overline{u'(t_o)^2}\ (t-t_o)^2$ represents the square of the typical
variation $\Delta_1 u$ of u in the interpolation interval:

$$\Delta u < \sqrt{\frac{\varepsilon}{2}}\Delta_1 u$$

In the case of a function of two variables, the same relation
is valid, $\Delta_1 u$ representing the typical variation of u in the
interpolation region.

With $\varepsilon'=0.2$, the interpolation region can be generally
assimilated to ellipses of sizes varying with height: about
0.7 to 1 hr in time, 3 km in height.   The corresponding Δ_1
is of the order of 15 m sec^{-1} at 90 km, and 20 m sec^{-1} at
100 km.   With $\varepsilon = \varepsilon'$, the corresponding error is

$$\Delta u \sim 4.5 \text{ m sec}^{-1} \text{ at 90 km}$$

$$\sim 6 \text{ m sec}^{-1} \text{ at 100 km} \tag{7.20}$$

Another error results from the error δu_y of each indivi-
dual wind measurement: it is of the order of $|\delta u_y|/\sqrt{n}$,
where n is the number of experimental data within the inter-
polation region.   Since n>3, and $|\delta u_y|$ has an upper limit given
by (7.15), the result is that the upper limit of this error is
equal or smaller to Δu, as given by (7.20).

Practical application.   The interpolation method described
above was applied to the data obtained at Garchy.   The computer
CAE 90-80 of the CNET was used, and the interpolated wind
$u(t_o, h_o)$ was computed for all the values of h_o between 80 and
110 km   and all the values of t_o every 10 minutes during the
recording run.   An example of the interpolated pattern so
obtained is shown in Fig. 7.16.

It appears that the results of this method can easily be
improved   by taking a few precautions:

(a) As the interpolation region R varies with t_o and h_o,

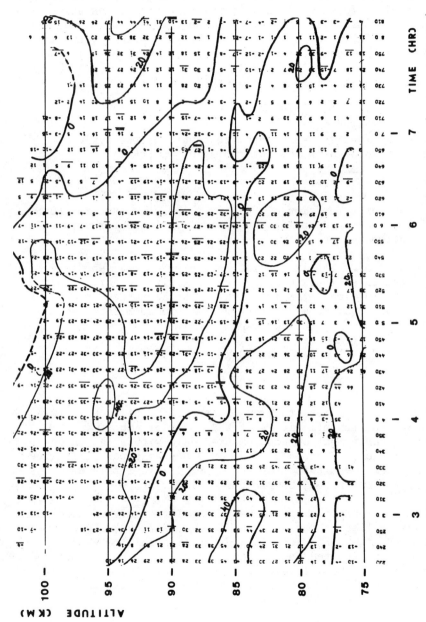

Fig. 7.16 Example of interpolated wind pattern (28-4-66). Underlined numbers correspond to the occurence of one meteor echo; the numbers underlined twice correspond to two echoes, or more.

a very small change of t_o and h_o can make one experimental point enter (or leave) R and then give rise to a discontinuity of the interpolated wind $u(t_o, h_o)$. All these small discontinuities look like a noise super-imposed on the wind pattern. This noise can easily be filtered out as follows: when applying the least-mean-square method used to determine the regression plane, a different weight is given to the experimental points: this weight is equal to 1 at the point t_o, h_o, and falls to zero at the edge of the interpolation region R.

(b) We have given above an upper limit of the r.m.s value Δu of the error but this error can exceed values much larger than Δu, with a small probability of this happening. In order to eliminate the values of u having the largest errors, it will be noticed that most of them correspond to extrapolations, i.e. the calculated $u(t_o, h_o)$ exceeds that greatest experimental value u_M of the wind within the interpolation region R, or is smaller than its smallest value u_m. We then only accept the interpolated values $u(t_o h_o)$ which satisfy

$$u_m \leqslant u(t_o, h_o) < u_M$$

(c) One could believe that the errors estimated in (7.20) are random (like a noise). On the contrary, as their prevailing term is caused by linearization in a small region of the (t, h) plane, these errors are mainly systematic, and the small-scale components are filtered out. More details concerning the filtering action of the interpolation method are given below, but we notice at once that such systematic errors can be greatly reduced, at least in one simple case. If a sinusoidal component of large period and wavelength is strongly prevailing (as the semi-diurnal tide above 95 km), the region R defined by (7.19) looks like a very elongated ellipse of large dimensions and eccentricity. A linear interpolation in such a region would nearly smooth out all the smaller-scale components. In this case, the interpolation is made in a region limited by an ellipse, smaller than R, of smaller eccentricity.

The interpolated wind patterns could be used for any kind of analysis, according to the physical phenomena that we want to bring out. Harmonic analysis is the most successful method, because the wind patterns always result from the superposition of a small number of sinusoidal waves, as we shall see in Sec. 7.6.

In order to discuss the best way to carry out this analysis,

we must anticipate a little the results that will be described
later in Sec. 7.6. As shown in this section, all the records
show clearly a strong and regular 12-hr tidal component, thus
confirming the results of previous experiments made at similar latitude (Greenhow, Neufeld, 1961; Kachtcheiev, Lysenko,
1961; Sprenger et al., 1967, etc). A prevailing component
also clearly exists, so that the whole wind power spectrum
S (f) can be considered as a sum of three components:

the semi-diurnal tide: $S_2(f) = u_2^2 \delta(f-f_2) + u_2^2 \delta(f+f_2)$
the prevailing wind: $S_0(f) = u_0^2 \delta(f)$
   other components: $S'(f)$ (7.21)

where δ is the Dirac function; u_2 the amplitude, $f_2 = 2$ day^{-1}
the frequency of the semi-diurnal tide; and u_0 the amplitude
of the prevailing wind. Here S (f) is defined for f negative by $S(-f) \equiv S(f)$.

The main cause of error is that the wind u_y (t) is known
only in a limited time interval, say (o, T), where T is the
duration of a continuous recording run. We can then compute
the estimated power spectrum

$$S_T (f) = \frac{1}{T} \left| \int_0^T u_y(t) \ e^{-2\pi jft} \ dt \right|^2 \qquad (7.22)$$

The discussion below leads to an order of magnitude of
this error and shows that it can be minimized by

(a) calculating the semi-diurnal tide with an integration
time T as an integer multiple of 12 hr, and substract it.

(b) calculating the spectrum of the remaining wind.

It can be shown (see, for instance, Blackmann and Tukey,
1958). The discussion summarized in the present section has
also been discussed in a more detailed way in appendix 2-5
of (Spizzichino, 1969a), where the error in the semi-diurnal
tide S_T (f) given by (7.22) is linked to the true power spectrum S (f) by

$$S_T (f) = T \int_{-\infty}^{+\infty} \left[\frac{\sin \ \{\pi T(f'-f)\}}{\pi T \ (f'-f)} \right]^2 S(f') \ df' \qquad (7.23)$$

When applying (7.23) with $f=f_2 = 2$ day^{-1}, each term of S(f)
given by (7.21) gives rise to a corresponding term of S_T (f):

(a) The term $u_2^2 \delta(f-f_2)$ gives rise to a term $T u_2^2$. If T is
large enough, this term is dominant. The equation

$$S (f) \simeq T u_2^2 \qquad (7.24)$$

yields an evaluation of u_2 which may then be applied as follows:

(b) The terms $u_2^2\delta(f+f_2)$ and $u_0^2\delta(f)$ give rise to terms which vanish, provided that the chosen time of integration T is always an integer multiple of 12 hr.

(c) The remaining spectrum S'(f) has most of its energy in higher-frequency terms (gravity waves; see Sec. 7.6). For these components, the ratio $\sin\{\pi T(f'-f_2)\}/ T(f'-f_2)$ is then small so that the corresponding term $S'_T(f)$ is also small. Furthermore $S'_T(f)$ can be estimated and then yields the error δu_2 on u_2 deduced by (7.24)

$$2\ u_2\ \delta u_2 = S'_T(f)$$

The error δu_2 has been so computed for all recording runs and for each altitude (all these individual values are given in Spizzichino, 1969a and b). It is generally of the order of 5 m sec^{-1} between 85 and 90 km.

Let us consider the error on the remaining spectrum S'(f). At least as a first approximation, the remaining spectrum as a sum of spectral lines, of a period of about one day (diurnal tide) or less than 8 hrs (gravity waves).

The amplitude u of any component of frequency f has an error δu caused by each of the other components of amplitude u' and frequency f'. After (7.23):

$$\frac{\delta u}{u} = \frac{1}{2}\left[\frac{\sin\{\pi T\ (f'-f)\}}{\pi T(f'-f)}\right]^2\ \frac{u'^2}{u^2} \tag{7.25}$$

When applied to the errors caused by the semi-diurnal tide $(f'=f_2)$, with $u_2=40$ m sec^{-1} and T=3 days, we find:

(a) for oscillations of a period of about 1 day (diurnal tide), or 6 to 8 hrs (long-period gravity waves)

$$\delta u/u \leqslant 0.1 \quad\text{for}\quad u = 10\text{ m sec}^{-1}$$

$$\leqslant 0.4 \quad\quad\quad 5\text{ m sec}^{-1}$$

(b) for oscillations of a period of less than 6 hrs (short-period gravity waves)

$$\delta u/u \leqslant 0.1 \quad\text{for}\quad u = 5\text{ m sec}^{-1}$$

The error δu can then be important in the first case. But let

us now substract the semi-diurnal tide given by (7.24) before
calculating the power spectrum: the error δu is then given by
(7.25) with $u' \sim \delta u_2$ instead of $u' = u_2$. In any case, we obtain

$$\frac{\delta u}{u} \leqslant 0.01 \qquad \text{for} \qquad u = 5 \text{ m sec}^{-1}$$

The error caused by the semi-diurnal tide is now quite negligible.

The same relation (7.25) can be applied to compute the
error over any component of the remaining spectrum, caused by
any other component, after having substracted the semi-diurnal
tide. The calculation has been made for all the components
observed at Garchy (Sec. 7.6) with typical values of their observed characteristics:

(a) a prevailing wind of the order of 20 m sec^{-1};
(b) a diurnal tide of amplitude 10 to 20 m sec^{-1}.
(c) gravity waves of amplitude 5 to 15 m sec^{-1}, of periods
1 to 8 hrs; the frequency difference between two of them being
at least of 1 day^{-1}. In all cases, it is found that

$$\frac{\delta u}{u} < 0.1$$

showing that the whole spectrum can be known with good accuracy.

As shown in Sec. 7.6 the remaining spectrum $S'(f)$ always
presents strong peaks (see also Fig. 7.35) of periods ranging
around 24 hrs (diurnal tide) or below 8 hrs (gravity waves).
Whenever such a peak appears, we have to determine whether it
represents a sinusoidal (or quasi-sinusoidal) individual oscillation, or a random maximum of a noise-like signal. In all
cases, the same tests can be applied to answer this question:

(a) If $S'(f)$ is indeed a noise-like spectrum, an estimation of its average value can be obtained by averaging the
values of the experimental spectrum for n successive values of
f at intervals of $1/2T$: In the case of a Gaussian wind, the
smoothed spectrum thus obtained must be distributed following
a χ^2 - law with n degrees of freedom (see, for instance, Blackmann and Tukey, 1958).

This smoothed spectrum has been systematically computed
for n=8 (in practical applications, we have in fact averaged
$S'(f)$ for 4 successive frequencies and 2 different altitudes):
in that case, the ratio of the 10% to the 90% level of its
distribution must theoretically be 3.7. Most of the peaks of
the smoothed experimental spectrum have a maximum value much

larger than 3.7 times the surrounding minima. Such peaks
then have a small probability of representing successive maxima
of a Gaussian noise. Furthermore, all the distributions of
such experimental spectra have a ratio in the 10% to 90% level
which is much larger than 3.7 (it varies between 4.5 and about
20, with a median value of 9), and always shows the broken
aspect of Fig. 7.17: a large number of small values and a
small number of much higher values corresponding to the peaks.

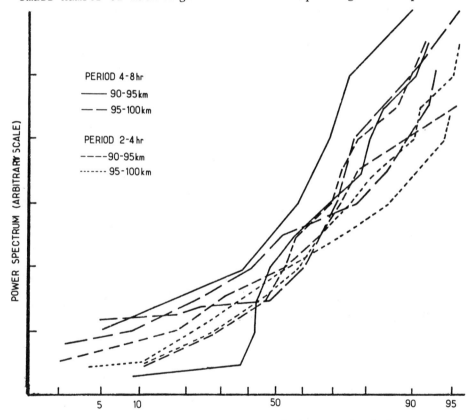

Fig. 7.17 Examples of repartitions of the zonal wind power
spectrum.

     (b) A peak often appears at the same frequency for all
(or most of) the altitudes; moreover, the phase of the corresponding oscillation is found to vary linearly with height,
as in the examples given in Section 6 (see Fig. 7.36). In
such cases, a much more convincing argument is obtained to
consider the peak as an individual oscillation.

     (c) This last characteristic suggests that most of these
peaks represent propagating waves. If this is so, its two-

dimensional power spectrum

$$S(f,k)=\frac{1}{t_o z_o}\lim_{\substack{t_o\to\infty\\z_o\to\infty}}\left[\int_{-t_o/2}^{+t_o/2}\int_{-z_o/2}^{+z_o/2}u_y(t,z)e^{2\pi j(kz\ -\ ft)}dt\ dz\right]^2$$

(where k is the reversal of the vertical wave-length) must show a peak.

As u (t, z) is known only in a time interval (0, T) and in a height interval (z, z + Z), we can compute the estimated power spectrum:

$$S_{TZ}(f,k)=\frac{1}{TZ}\left[\int_0^T\int_{z_o}^{z_o+Z}u_y(t,z)e^{2\pi j(kz\ -\ ft)}dt\ dz\right]^2$$

It can be shown that S_{TZ} is linked to the true power spectrum S(f, k) by a relation similar to (22):

$$S_{TZ}(f,k)=TZ\int_{-\infty}^{+\infty}\int_{-\infty}^{+\infty}\left[\frac{\sin\{\pi T(f-f')\}}{\pi T(f-f')}\ \frac{\sin\{\pi Z(k-k')\}}{\pi Z(k-k')}\right]^2 S(f'k')df'dk'$$

$$(7.26)$$

The advantage of using the two-dimensional spectrum appears when applying (7.25) and (7.26) to a signal containing a sinusoidal wave of amplitude V_o to a random noise of spectrum B (f, k)

$$S(f,k)=V_o^2\ \delta(f-f_o)+B(f,k)$$

The one and two-dimensional spectra, respectively, given by (7.25) and (7.26) are both the sum of two terms:

(a) a peak, of maximum amplitude V_o^2
(b) a noise-like signal, equal to

$$\int_{-\infty}^{+\infty}\int_{-\infty}^{+\infty}\left[\frac{\sin\{\pi T(f-f')\}}{\pi T(f-f')}\right]^2 B(f',k')df'dk'$$

for the one-dimensional spectrum   and to

$$\int_{-\infty}^{+\infty}\int_{-\infty}^{+\infty}\left[\frac{\sin\{\pi T(f-f')\}}{\pi T(f-f')}\ \frac{\sin\{\pi Z(k-k')\}}{\pi Z(k-k')}\right]^2 B(f',k')df'dk'$$

for the two-dimensional spectrum. This noise is obviously
much smaller in the last case; a numerical application (for
more details, see (Spizzichino, 1969 a , appendix 2.5) would
show that it is about 6 times smaller. If the wind pattern
contains one (or a few) propagating wave, a two-dimensional
spectrum analysis will make it appear more significant.

Such peaks frequently appear in the two-dimensional
spectrum deduced from the Garchy observations. Examples are
given below in Section 6 (Figures 7.30-7.32) where this kind
of analysis will help us to study the diurnal tide. The amplitude of these peaks is generally 10 to 50 times larger than
the surrounding noise. Furthermore, the variations of the
experimental spectrum S_{TZ} (f,k) near these maxima closely
coincide with those theoretically deduced from (7.24) for a
single wave (i.e., S(f, z) being a Dirac function). An example
is given in Fig 7.18 for a particular wave of a period of
16 hrs observed in April 1966.

In brief, three possible tests (a), (b), (c) have been
described in order to recognize the waves contained in the
wind pattern. The description of gravity and tidal waves
given in the following section will be restricted to waves
satisfying all these three tests.

In order to test the validity and the limits of both the
interpolation method and the spectrum analysis, a series of
trials has been made: a wind profile is supposedly known, and
is a single sinusoidal wave

$$u(t, z) = u_o \cos (\omega t - kz)$$

or a sum of such waves. A random series of t and z values is
chosen: t_i, z_i, with i = 1 to n. The corresponding $u_i = u$
(t_i, z_i) is computed from the above formula; the series of
values t_i, z_i, u_i are then taken as experimental data and
analyzed by our interpolation program. A Fourier analysis
of the resulting wind pattern is then made, and the obtained
spectrum is compared with the spectrum which has been introduced.

This test was made for different rates of experimental
data. It showed that:

1) The interpolation method acts like a filter, cutting
off the components of period smaller than 1 or 2 hrs. Its
response (Fig. 7.19) depends on the hourly rate of data ρ
but shows small variations for $\rho=15$ to 35 (practical values
of the hourly rates of meteor echoes)

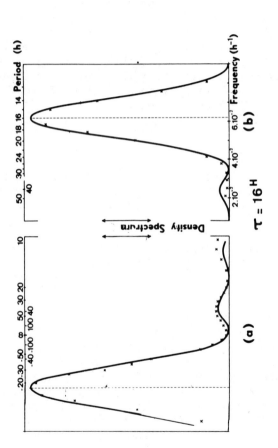

Fig. 7.18 Two-dimensional spectrum S(f, k), Garchy, April 27-28, 1966. The variations of S(f, k) are represented along a line of constant k (wavelength: 19 km). The experimental points (crosses) are compared to the theoretical estimated power spectrum corresponding to a spectral line.

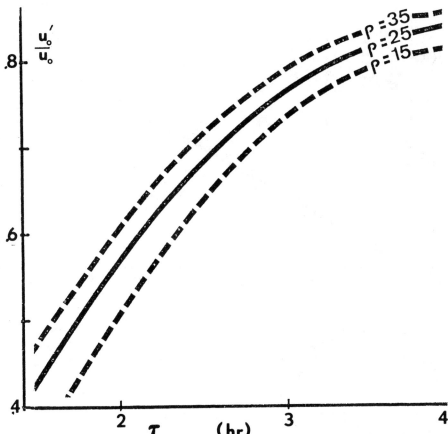

Fig. 7.19 Error caused by the method of analysis. This method
acts like a filter without phase distorsion, but with an
attenuation u_o'/u_o varying with the pediod τ. The attenuation
depends on the hourly rate of echo ρ, but for $\rho > 15$ it presents only small variations around a mean value.

2) This filter has no phase distorsion. In particular,
the phase variation with respect to height is correctly restituted, even for a small rate of experimental data (Fig. 7.20).
The vertical propagation of a gravity wave must then be observed and its vertical wavelength is obtained with a very
good accuracy.

3) The intermodulation effects are small. If the original wind pattern is a single sinusoidal wave of amplitude
1, the obtained interpolated pattern does not contain any
other component of amplitude larger than 2.10^{-2} (Fig. 7.21).

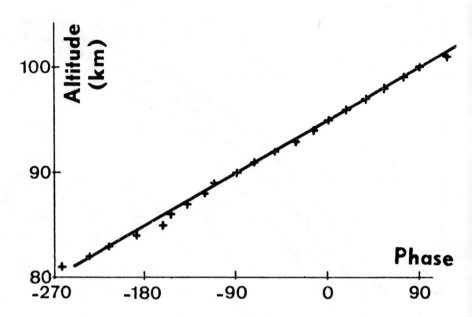

Fig. 7.20 Result of the analysis of a sinusoidal wind profile τ=2 hr and wavelength λ=20 km, with ρ=12 detected
echoes per hr. After interpolation and harmonic analysis,
the phase variation of this oscillation is restituted correctly.

7.6 Experimental Results

The experimental results described below are those of a
first series of 10 recording runs made at Garchy (47°N, 3°E)
from Sept. 1965 to Sept. 1966 and were obtained under much
less favorable conditions than in the following years, and
they have to be considered as preliminary ones.

(a) The duration of each recording run was 2 to 3 days
(they now last about 10 consecutive days in 1970).
(b) The photographic recording method was used, thus
resulting in an important loss of data (Sec. 7.4), but magnetic tape records have been used since 1969).
(c) Only the east-west component of the wind has been
measured.

However, many new results were obtained, mainly owing
to the small height definition of the equipment. Some of these
have already been published elsewhere (Revah, 1968; Spizzichino, 1969 b, 1970) and will be briefly summarized below.

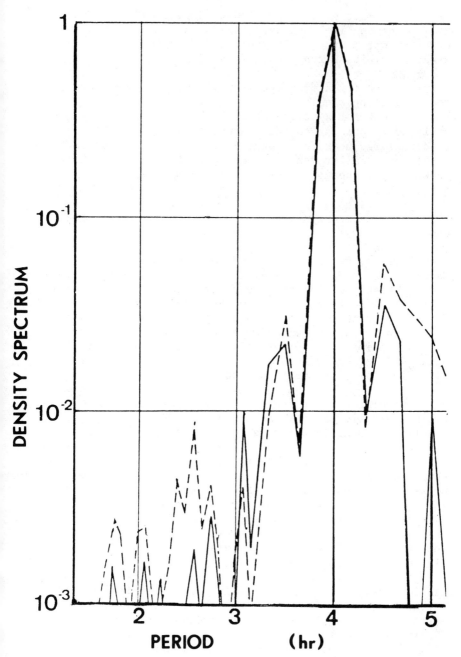

Fig. 7.21 Power spectrum obtained from the analysis of a wind profile of period 4 hr , with $\rho=8$ detected echoes per hour (dashed line) and $\rho=32$ echoes per hour (full line).

An example of interpolated wind pattern is given in Fig. 7.22. The zonal wind always varies in a complex way, but its power spectrum at all altitudes and for all the recording runs always exhibits strong peaks (Fig. 7.23):

(a) a peak for a frequency equal to zero, corresponding to the prevailing wind;
(b) a peak for a period of 24 hr   (or about 24 hr)   probably corresponding to the diurnal tide;
(c) a large peak for a period of 12 hr   (semi-diurnal tide);
(d) many smaller peaks for periods of less than 8 hr   as shown below, they probably result from a superposition of gravity waves.

The mean energy (per mass unit) of these four zonal wind components is given in Fig. 7.24 for each altitude.

Firstly, we shall describe some results obtained at Garchy concerning the prevailing wind and the semi-diurnal tide. These two wind components are found to vary slowly with height so that a small height resolution is not needed for studying them. They have also been observed from other experiments, many of which have been carried out in Europe. Our main point of interest will be to compare our results with those of other European stations, in order to obtain a synoptic view of these phenomena.

The diurnal tide and the gravity waves have been found to exhibit drastic variations with height, and the Garchy radar enables us to discover new aspects concerning them.

The prevailing wind is the wind component of frequency equal to zero. It is computed as described below and then represents the mean value of the wind during the recording run.

The variations of the prevailing zonal wind at Garchy are given for all the recording runs with respect to altitude in Fig. 7.25. The error intervals are not plotted in this figure. The error caused by the small-scale components is always negligible and that caused by the semi-diurnal tide is eliminated by substracting it. The remaining error is then of the order of a few m sec^{-1}, with certain exceptions for months where strong long-period oscillations appeared. The fluctuations of the computed prevailing wind with respect to altitude in February and April 1966 probably result from that effect.

The prevailing wind is found to exhibit systematic seasonal variations similar to those observed in other European

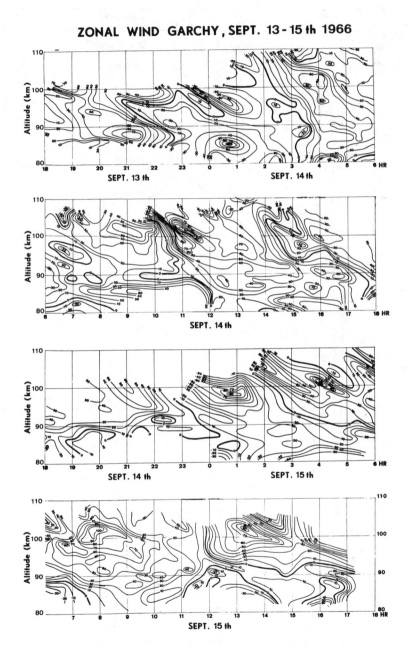

Fig. 7.22 An example of interpolated wind pattern at Garchy
(Sept. 13-15, 1966).

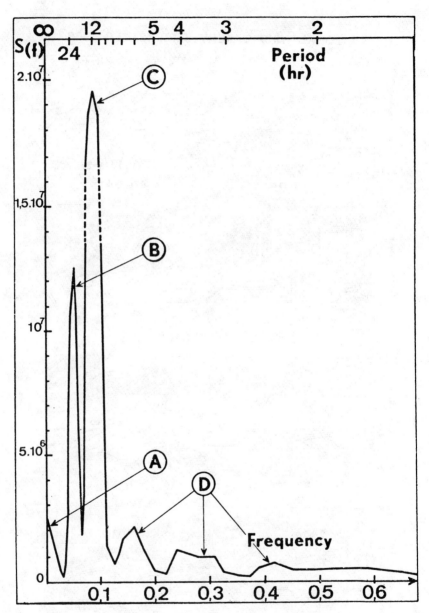

Fig. 7.23 Power Spectrum of the zonal wind, Garchy, 29 March-
Apr. 1, 1966. The power spectrum S (f) is given in m^2 s^{-1} in
ordinates with respect to frequency (in abscisse, in hr^{-1};
upper scale: periods, in hr). It delineates the 4 main components of the wind in the upper atmosphere: prevailing wind
(A), diurnal tide (B), semi-diurnal tide (C), gravity waves
(D).

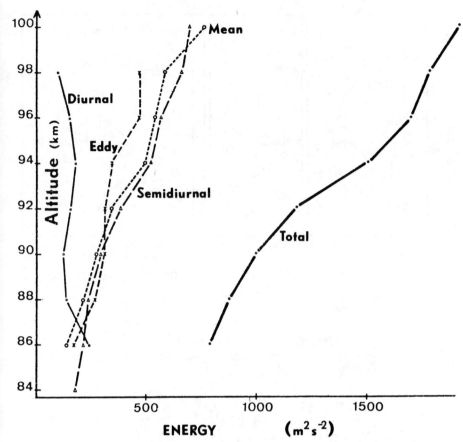

Fig. 7.24 Mean energy per mass unit of the zonal wind components, computed over all the recording runs of 1965-66.

stations of similar latitude (Fig. 7.26). This is true in the
short period variations as well as in the zonal wind. The
spectrum analysis of the zonal wind observed at Garchy always
shows a strong oscillation of period 12 hrs.

Some examples of its phase and amplitude variations are
shown in Figures 7.27 a, b, c, d. Its amplitude generally
increases with height, thus agreeing with the classical
theory of tides. Its phase either does not exhibit significant variations or slowly increases with height. This last
case would correspond to a wave of downward phase velocity and
of large vertical wavelength (>80 km), as expected for the
theoretical S_2^2 mode.

The phase and the amplitude of the semi-diurnal tide are
found to exhibit systematic seasonal variations which are

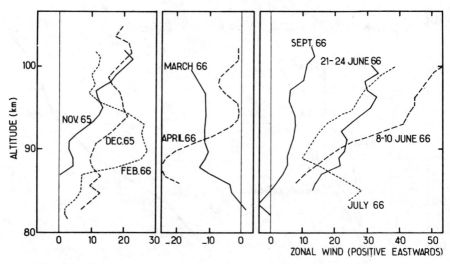

Fig. 7.25 Zonal prevailing wind at Garchy (1965-1966).

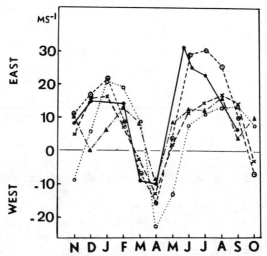

Fig. 7.26 Comparison of the seasonal variations of the zonal
prevailing wind observed at Garchy (1965-66, solid curve),
Jodrell-Bank (1953-58, dash-cross curve), Kuhlungsborn (1965,
dash-dot curve), Hkarkov (1964-65, dash-dot-triangle curve)
and Obninsk (1964-65, dot curve).

shown in Fig. 7.28 and compared with those of other European
stations for different years. The seasonal variations of the
phase are rather similar for all these stations; those of the
amplitude exhibit stronger differences, possibly caused by
their rapid variation with respect to the altitude. As the
method of observation varies from one station to another, the

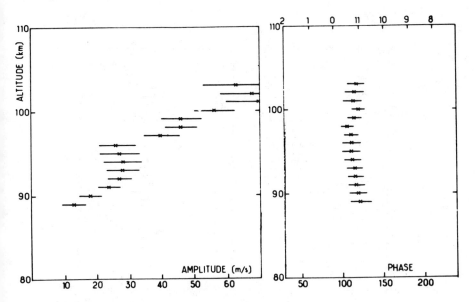

Fig. 7.27a   Amplitude and phase of the semi-diurnal tide observed
at Garchy (Nov. 16-17, 1965).

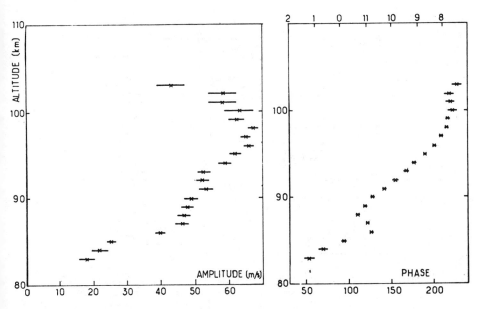

Fig. 7.27b   Amplitude and phase of the semi-diurnal tide observed
at Garchy (Feb. 22-24, 1966),

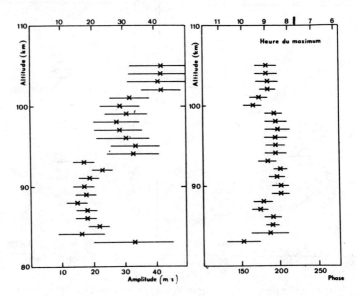

Fig. 7.27c Amplitude and phase of the semi-diurnal tide observed
at Garchy (July 19-22, 1966).

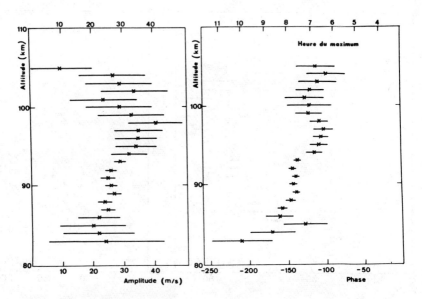

Fig. 7.27d Amplitude and phase of the semi-diurnal tide observed
at Garchy (Sept. 13-15, 1966).

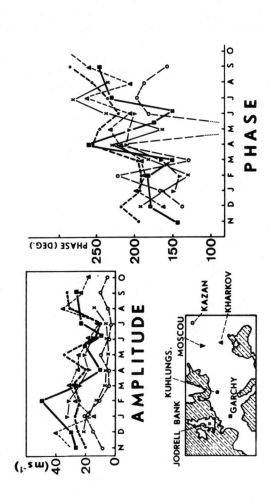

Fig. 7.28 Comparison of the seasonal variations of the zonal semi-diurnal tide observed at different European stations: Garchy, Kuhlungborn, Jodrell Bank, Kharkow, Moscow and Kazan.

mean altitude of the observation can be slightly different.
This factor can be expected to be of small information in the
analysis. Let us remember that the diurnal oscillation is
theoretically expected (Kato, 1966; Lindzen, 1966) to be a sum
of two kinds of mode:

(a) The propagating modes which propagate upwards like a
progressive wave. The vertical wavelength is about 25 km
for the first of these modes and smaller for the following ones.
Their energy density (per volume unit) should not vary with
height, provided that viscous damping can be neglected.
(b) The evanescent modes; in the absence of energy source,
their energy density decreases exponentially with height and
their phase does not rotate.

Up to now, the observations made at meteoric heights
(80 to 105 km) did not give a clear confirmation of the theoretical analysis given above. The amplitude and phase of the
diurnal tide were found to present irregular variations (Greenhow, Neufeld, 1961; Elford, 1959), and did not seem to follow
any simple law. However, rocket measurements made at higher
or at lower altitudes yielded a more coherent description,
with a predominance of the first propagating mode (Hines, 1966;
Miers, 1965; Beyers et al. 1965, 1966; Reed et al., 1966;
Reed and Oard, 1969).

It sometimes happens that the Fourier analysis of the
winds measured at Garchy shows a 24 hr oscillation and a
phase increasing linearly with height, as shown in Fig. 7.29.
The observed tidal oscillation then has a downward phase
velocity and a vertical wavelength of about 20 km, as expected
for the first propagating mode.

However, in most cases, the phase of the diurnal oscillations was found to exhibit complex variations which could
suggest the superposition of many modes of different vertical
wavelength.

The best way of studying the diurnal tide is to compute
the two-dimensional power spectrum S(f, k). Although a
smoothed spectrum can only be obtained, its definition is
small enough to separate the first three propagating modes
of the diurnal tide.

In general, the spectra S(f, k) so obtained present a
small number of maxima, for periods of 24 hr or near this
value. The variations of the experimental spectra S(f, k)
near each maximum closely coincide with those of spectrum obtained theoretically for a single propagating wave.

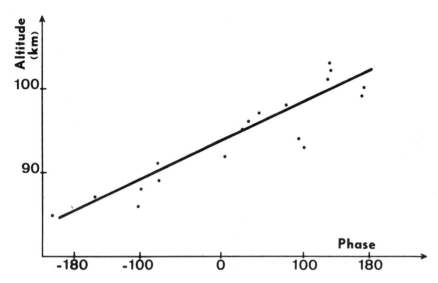

Fig. 7.29 Phase of the diurnal oscillation observed at Garchy, 14-16 Dec. 1965. In that particular case, an oscillation of period 24 hr and vertical wavelength 20 km is observed, as expected from theory.

Figures 7.30, 7.31 and 7.32 show some examples of typical observed spectra. The spectrum given in Fig. 7.30 (March 1966) presents two maxima, each for a period of 24 hrs. The stronger one has a vertical wavelength of about 25 km and a downward phase velocity (exactly like the first positive mode described by theory); the other has nearly the same wavelength but an upward phase velocity.

In many cases (Fig. 7.31), a similar spectrum is obtained, but the period of the dominant oscillation is significantly different from 24 hrs. Waves of periods of 16 to 30 hrs. are found, without preference for the value of 24 hr. The mean spectrum calculated over all the records presents a very smooth maximum for periods near 24 hr. Moreover, all these oscillations have vertical wavelengths of the order of 20 to 40 km. We then have no reason to consider the 24 hr-oscillations and the other ones as two fundamentally different phenomena. From a physical point of view, a 16 hr-oscillation is likely to represent a diurnal tide whose phase has been strongly shifted from one day to the next.

In all these records, no oscillation has been found with a vertical wavelength smaller than 15 km (as can be expected for the theoretical second propagating mode and the following ones).

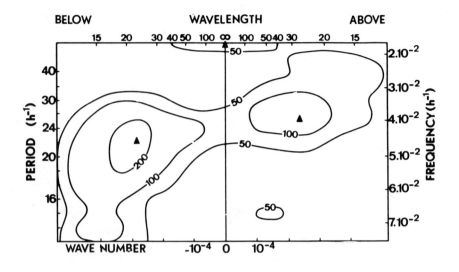

Fig. 7.30 Two-dimensional power spectrum of the zonal wind,
Garchy Mar. 29-Apr. 1 1966.

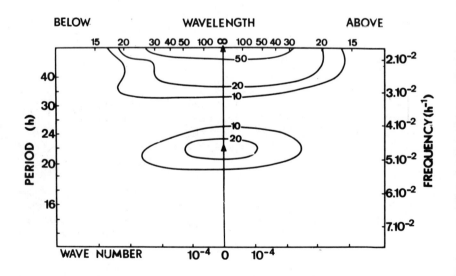

Fig. 7.31 Two-dimensional power spectrum of the zonal wind,
Garchy Jun. 21-24, 1966.

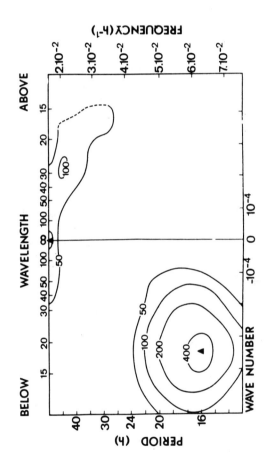

Fig. 7.32   Two-dimensional power spectrum of the zonal wind, Garchy Apr. 27-28, 1966.

A few recording runs, made in June and July (Fig. 7.32) showed a diurnal oscillation with an infinite vertical wavelength. The phase does not change with height, as in the theoretical evanescent modes.

The difference between the diurnal oscillations observed in June and July, on the one hand, and during the remaining part of the year on the other hand, clearly appears in Fig. 7.33 where we have plotted all the maxima of the two-dimensional power spectrum observed during the recording runs of 1965-66.

The mean energy density (per volume unit) of these diurnal (or, rather quasi-diurnal) oscillations has been computed separately for June and July and for the remaining months. The results are shown in Fig. 7.34.

The energy density is found to decrease very rapidly with increasing height in June and July, thus confirming our hypothesis of prevailing evanescent modes during this part of the year.

If we try to represent its variations by an exponential law, $W \propto \exp(-az)$, the best fit is obtained for $a=0.26$ km^{-1}. This value would correspond to a theoretical equivalent height $h=-1.9$ km (which is near to the theoretical value $h=-1.75$ obtained for the second evanescent mode).

The result obtained during the other months is more surprising. The energy density of the quasi-diurnal oscillation is found to decrease with increasing height (although much more slowly than in June and July). The viscous damping at meteoric heights is not strong enough to explain an energy loss by a factor of 5 between 88 and 98 km (Fig. 7.34) for waves having vertical wavelengths of 20 to 40 km.

In conclusion, the wind measurements made at Garchy during 1965-66 produced evidence that the diurnal tide results from the superposition of evanescent and propagating modes, but some anomalies can be observed in the properties of the first propagating mode:

(a) strong phase shifts (it is more correct to speak of a quasi-diurnal oscillation than of a diurnal tide);
(b) existence of waves with an upward phase shift (like reflected waves);
(c) great loss of energy at meteoric heights.

An attempt has been made elsewhere to explain these

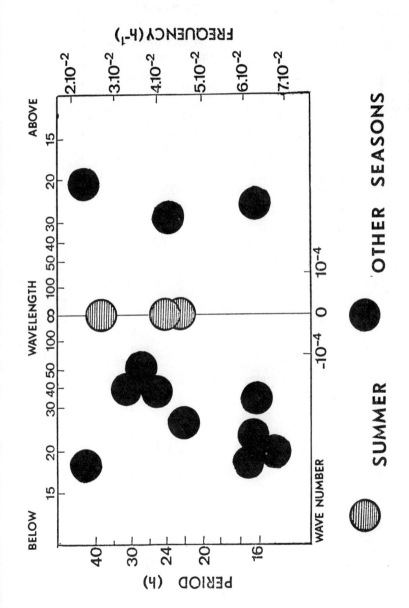

Fig. 7.33 Maxima of the two-dimensional power spectrum observed at Garchy during the recording runs of 1965-66.

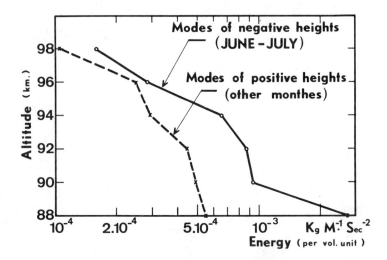

Fig. 7.34 Variations of the energy density (per volume unit)
vs height at Garchy. In June and July, it decreases with
increasing height as expected for the evanescent modes.
During the remaining part of the year, it decreases more slowly.

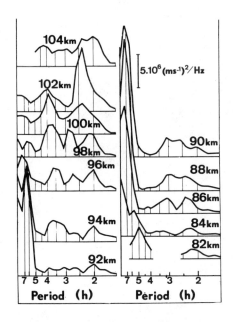

Fig. 7.35 Power spectrum of the zonal small-scale wind,
Garchy, Dec. 14-15, 1965.

anomalies involving interactions between the diurnal tide and the other components of the wind (Spizzichino, 1969 b and 1970).

The power spectrum of the wind always exhibits strong peaks for periods ranging between 2 and 8 hours, as shown in the examples of Fig. 7.35. We shall consider as significant only those peaks which satisfy the three tests described in the last section. For these peaks:

(a) The phase always increases linearly with height, as shown in the examples of Fig. 7.36 (it was found to decrease linearly in one single case); like waves, these oscillations then propagate with a constant downward phase velocity ranging between 1 and 20 km per hour: they could represent gravity waves the energy of which is propagated upwards (Hines, 1960).

(b) Most of these waves have a vertical wavelength ranging between 15 and 30 km (Fig. 7.36).

(c) The averaged power spectrum of these oscillations, calculated for all the recording runs of 1965-66, does not exhibit any more significant peaks; at each altitude, it decreases with increasing frequency f proportionally to $f^{-\alpha}$, with

α= 0.82 at 90 km

α= 0.47 at 100 km

(Fig. 7.37); the coefficient α regularly decreases with increasing height.

(d) Most of these oscillations have an amplitude varying with altitude, as shown in Fig. 7.38; the height interval between two successive maxima and minima is about one half of the vertical wavelength deduced from the phase variations. These amplitude variations probably result from the interference of two progressive waves, one of them having a downward phase velocity and the other having an upward phase velocity and a smaller amplitude; the existence of the latter could be explained by partial reflections of gravity waves at levels slightly higher than meteoric heights; the reflection coefficient R can be deduced from the variation law of the resulting field

$$u(z)=Ue^{\frac{z}{2H}}\{1+R^2+ 2R \sin (2kz + \alpha)\}$$

where k is the vertical wave number and H the height scale;

the constants R, U, α and H can be computed by a least-mean-square method; R is found to vary between 0.1 and 0.7 (mean value: 0.3) (Revah, 1969 and 1970).

(e) The values of H so obtained are generally smaller than expected; one half of them are smaller than 5 km and

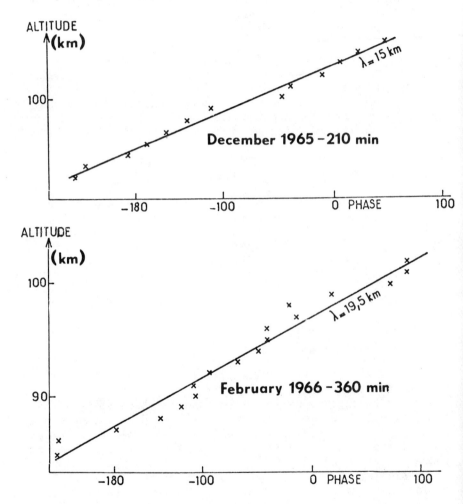

Fig. 7.36 Examples of phase variation vs. height for two particular peaks of the small-scale wind power spectrum.

would then correspond to temperatures lower than 150°K (Fig. 7.39); as such temperatures are not considered to be usual between 80 and 110, we must conclude that the values of H so computed are smaller than the true ones; the amplitude of the

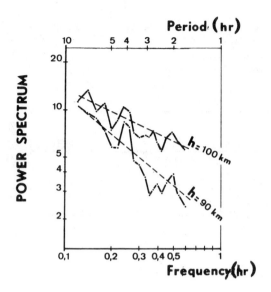

Fig. 7.37 Mean power spectrum of the zonal wind at 90 and 100 km, computed over all the recording runs of 1965-66. The spectral density decreases with increasing frequency f as $f^{-0.82}$ at 90 km, and as $f^{-0.47}$ at 100 km.

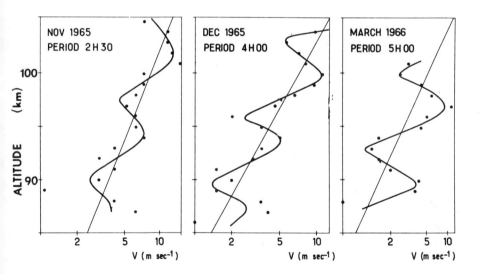

Fig. 7.38 Examples of variations of the gravity wave amplitude with height, showing the superposition of an exponential increase, and of sinusoidal fluctuations.

upward propagating waves then increases more rapidly than $e^{z/2H}$, and their energy increases with height; to explain this anomaly, a mechanism has to be found in which an energy is given to the wave (Spizzichino, 1969 b, 1970).

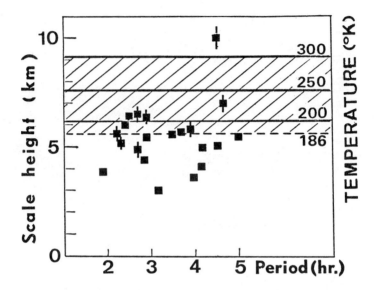

Fig. 7.39 Values of the scale height H deduced from the increase of the amplitudes of the gravity waves observed at Garchy. One half of them are smaller than 5 km and would then correspond to temperature lower than 150°K.

7.7 Conclusions

Although the series of recording runs made in 1965-66 did not use all the possibilities of the equipment described above, they showed many theoretically expected phenomena: propagation of gravity waves and the existence of propagating and evanescent modes of the diurnal tide. However, they also showed many anomalies of these waves: energy increase, energy loss and phase instability of the diurnal tide. These anomalies could be explained by taking account of the energy exchanges between these waves as discussed in other publications (Spizzichino, 1969 b, and 1970).

7.8 References

Bean, R. B., and G. D. Thayer, 1959. "Models of the Atmospheric Radio Refractive Index," Proceedings of Institute of Radio Engineers, 47, 5, 740-755.

Beyers, N. J., B. T. Miers, 1965. "Diurnal Temperature Changes in the Atmosphere between 30 and 60 km over White Sands Missile Range", Journal of Atmospheric Sciences, 22, 262-266.

Beyers, N. J., B. T. Miers, and R. J. Reed, 1966. "Diurnal Tidal Motions near the Stratopause during 48 hr. at White Sands Missile Range," Journal of Atmospheric Sciences, 23, 325-33.

Blackman, R. B., J. W. Tukey, 1958. Bell System Technical Journal 37, 185-282 et 485-570.

Delcourt, J., I. Revah, A. Spizzichino, 1963. "Etude Goniometrique de la Haute Atmosphere," Space Research IV, 182.

Edwards, H. D., et al., 1963. "Upper Atmosphere Wind Measurements Determined from Twelve Rocket Experiments," Journal of Geophysical Research, 68, 3021.

Elford, W. G., 1959. "A Study of Winds between 80 and 100 km in Medium Latitudes," Planetary and Space Science, 1, 94-101.

Elford, W. G., 1964. "Seasonal and Diurnal Variations of Winds at 90 km", (Abstract), Transactions of the American Geophysical Union, 45, 60.

Glass, M., I. Revah, A. Spizzichino, 1969. "Radars Meteoriques Mobiles pour l'etude des Vents dans la Haute Atmosphere," Note Tech. EST/RSR/47.

Greenhow, J. S., 1954. "Systematic Wind Measurements at Altitudes of 80 to 100 km using Radio Echoes from Meteor Trails", Phil. Magazine, 45, 471.

Greenhow, J. S., 1952. "Characteristics of Radio Echoes from Meteor Trails", III. "The Behaviour of Electron Trails after Formation", Proceedings of the Physical Society of London, B 65, 169.

Greenhow, J. S., 1950. "The Fluctuation and Fading of Radio Echo, from Meteor Trails", Phil. Magazine 41, 682.

Greenhow, J. S., E. L. Nuefeld, 1955. "Diurnal and Seasonal Wind Variation in the Upper Atmosphere", Phil. Magazine 46, 549-562.

Greenhow, J. S., E. L. Neufeld, 1955. "The Diffusion of Ionized Meteor Trails in the Upper Atmosphere", Journal of Atmospheric and Terrestrial Physics, 6, 133-140.

Greenhow, J. S., E. L. Neufeld, 1956. "The Height Variation of Upper Atmosphere Winds", Phil Magazine, 1, 1157-1171.

Greenhow, J. S., E. L. Neufeld, 1961. "Wind in the Upper Atmosphere", Quarterly Journal of the Royal Meteorological Society, 87, 472-489.

Hines, C. O., 1960. "Internal Atmospheric Gravity Waves at Ionospheric Heights", Canadian Journal of Physics, 38, 1441-1481.

Hines, C. O., 1966. "Diurnal Tide in the Upper Atmosphere", Journal of Geophysical Research, 71, 1453-1459.

Kachtcheiev, B. L., I. A. Lysenko, 1961. "Investigation of Atmospheric Circulation at the Height of 80-120 km" (en russe), Ionospheric Research 9, Academy URSS.

Kato, S., 1966. "Diurnal Atmospheric Oscillations". 1. Eigenvalues and Hough Functions, Journal Geophysical Research, 71, 3201-3209. 2. "Thermal Excitation in the Upper Atmosphere", Journal Geophysical Research 71, 3210-3214.

Kent, G. S., 1960. "The Fading of Radio Waves Reflected Obliquely from Meteor Trails", Journal of Atmospheric Research, 19, 272.

Lindzen, R. S., 1966. "On the Theory of the Diurnal Tide", Monthly Weather Review, 94, 295-301.

McKinley, D. W. R., 1961. "Meteor Science and Engineering", McGraw-Hill Book Co. Inc.

Miers, B. T., 1965. "Wind Oscillations Between 30 and 60 km over White Sands Missile Range", Journal of Atmospheric Science, 22, 382-387.

Manning, L. A., 1959. "Air Motion and the Fading Diversity, and Aspect Sensitivity of Meteoric Echoes", Journal of Geophysical Research, 64, 1415-1425.

Manning, L. A., O. G. Villard, A. M. Peterson, 1950. "Meteoric Echo Study of Upper Atmosphere Winds", Proceedings Institute for Radio Engineers, 38, 877-883.

Reed, R. J., M. J. Oard, 1969. "A Comparison of Observed and Theoretical Tidal Motions Between 30 and 60 km., Monthly Weather Review, 97, 456-459.

Reed, R. J., D. J. McKenzee, J. C. Vyverberg, 1966. "Diurnal Tidal Motions Between 30 and 60 km in Summer," Journal of Atmospheric Science 23, 416-423.

Revah, I., 1968. "Etude des vents de petite echelle observes au moyen des trainees meteoriques"(these de Docteur es Sciences.)

Revah, I., 1970. "Partial Reflections of Gravity Waves Observed by Means of a Meteoric Radar", Journal of Atmospheric and Terrestrial Research, to be published.

Revah, I., 1969. "Etude des vests de petite echelle observes au moyen des trainees meteoriques,"Annals of Geophysics, t 25, Fasc. I.

Revah, I., A. Spizzichino, C. Taieb, 1962. "Etude au sol de la basse ionosphere". Note technique CDS/GRI/10.

Revah, I., A. Spizzichino, M. Massebeuf, 1966. "Premiere Campagne de Mesure des Vents Ionospheriques avec le Radar Meteorique de Garchy", Note technique CDS-37, Paris.

Revah, I., A. Spizzichino, C. Taieb, 1963. "Etude des cisaillements de vents dans la basse ionosphere par l'observation radioelectrique des trainees meteoriques", I Etude theorique, Annals of Geophysics, 19, 43-50.

Revah, I., A. Spizzichino, 1964. "Etude des cisaillements de vents dans la basse ionosphere par l'observation radioelectrique des trainees meteoriques," II Etude experimentale, Annals of Geophysics, 20m 248-260.

Roper, R. G., W. G. Elford, 1965. "Meteor Winds Measured at Adelaide (35 S), 1961; NASA Technical Publication, X-650-65-220, Goddard Space Flight Center, Greenbelt, Maryland.

Schulkin, M., 1952. P.I.R.E., mai.

Spizzichino, A., 1969. "Etude des interactions entre les differentes composantes du vent dans la haute atmosphere" (these de doctorat d'Etat.).

Spizzichino, A., 1969. "Etude experimentale des vents dans la haute atmosphere." Annals of Geophysics, 25, 5-28.

Spizzichino, A., 1969. "Quelques donnees theoriques sur la propagation des ondes atmospheriques." Annals of Geophysics, 25, 755-771.

Spizzichino, A., 1969. "Theorie des interactions non lineaires entre les ondes atmospheriques." Annals of Geophysics, 25, 773-783.

Spizzichino, A., 1970. "Etude des interactions entre la maree diurne et les ondes de gravite." Annals of Geophysics 26, 9-12.

Spizzichino, A., 1970. "Autres applications de la theorie des interactions non lineaires; Conclusion". Annals of Geophysics, 26, 25-34.

Spizzichino, A., J. Delcourt, A. Giraud, I. Revah, 1965. "A New Type of Continuous Wave Radar for the Observation of Meteor Trails", Proceedings of Institute of Electronics and Electrical Engineers, 53, 1084-1086.

Sprenger, K., R. Schminder, 1967. "Results of Then Years Ionospheric Drift Measurements in the 1. f. Range", Journal of Atmospheric and Terrestrial Physics, 29, 183-199.

CHAPTER 8

RADIO METEOR WINDS IN THE
SOUTHERN HEMISPHERE

R. G. Roper *

School of Aerospace Engineering,
Georgia Institute of Technology,
Atlanta, Georgia

8.1 Introduction

The radio meteor system used at *Adelaide, Australia,*
(35°S, 139°E) for measuring the velocity of drift and position
of meteor trails was initiated by Professor L. G. H. Huxley
in 1950, and has been described by Robertson et al.(1953); an
updated system, with three receiving sites, which has also
been used to determine meteor orbits and atmospheric turbulence
from individual trail shears since 1960, has been detailed by
Weiss and Elford (1963). Continuous wave radiation (1.5 kw)
on a frequency of 26.773 MHz is emitted vertically within a
cone of semi-angle 60°, and a meteor trail suitably oriented
within this cone reflects radiation back to an antenna array
at the main receiving site, by means of which the direction
of the reflection point can be determined. The component of
the velocity of trail drift in the direction of observation
is derived from the doppler frequency change of the reflected
wave, while the range of the reflection point is found by the
radar technique, using a 65 kilowatt peak pulse output transmitter on a frequency of 27.540 MH, and a broadband receiver
also located at the main receiving site 22.9 kilometers north
of the transmitters. The four outstations at present in use
enable the drifts of separate points on the trail, up to 20
kilometers apart, to be measured.

While the separation between reflecting segments on any
given trail can be measured to within 100 meters, the height

---

* Formerly of the Department of Physics, University of Adelaide, Australia.

of the reflecting center appropriate to the main receiving
site can be measured to an accuracy of ± 2 kilometers, which
adequately resolves the prevailing wind and the dominant tidal
components over the height range 75 to 105 kilometers.

Some effort is being directed toward automation of the
data acquisition but, at the moment, echo information is still
being recorded in analog form on 35 mm film. Semi-automatic
film reading produces one punched card per echo, and these
cards are computer processed to produce line of sight drift
(magnitude and direction), reflection point height, and local
mean solar time of occurrence for each echo.

8.2 Observational Technique

Two transmitters are used (Fig. 8.1). *Continuous wave
radiation* (1.5 kw) on a frequency of 26.773 MHz is emitted
vertically in a cone of semi-angle 60°, and a meteor trail
suitably oriented within this cone (Fig. 8.2) reflects radiation back to an antenna array (Fig. 8.3) at the main receiving site, by means of which the direction of the reflection
point can be determined. The component of the velocity of
trail drift in the direction of observation is derived from
the doppler frequency change of the reflected wave, while the
range of the reflection point is found by the radar technique,
using a 65 kw peak pulse output transmitter on a frequency of
27.540 MHz, and a broadband receiver also located at the main
receiving site 22.9 km north of the transmitters. A block
diagram of the receiving equipment, including details of the
remote receiving site equipment, is shown in Fig. 8.4. The
five outstations at present in use enable the drifts of separate points on the trail, up to 20 km apart, to be determined.

While the separation between reflecting segments on any
given trail can be measured to within 100 meters, the height
of the reflecting center appropriate to the main receiving
site can be measured to an accuracy of only ± 2 km. This
resolution adequately resolves the prevailing wind and the
dominant tidal components over the height range 75 to 105 km.

Some effort is being directed toward automation of the
data acquisition, but, at the moment, echo information is
still being recorded in analog form on 35 mm film. A section
of one of these films, covering two echoes, is shown in
Fig. 8.5. The top trace is a conventional A scan radar presentation, with 20 km rangemarkers. The bottom traces, below
the echo identification numbers, are the outputs of the two
direction finding receivers, which are switched between three

RADAR TRANSMITTER

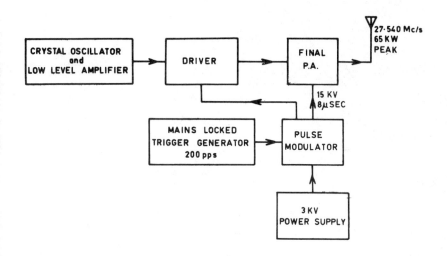

C.W. TRANSMITTER

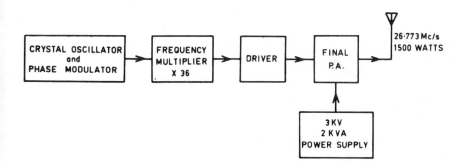

Fig. 8.1   System block diagram

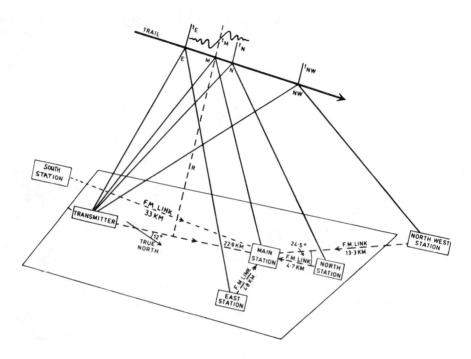

Fig. 8.2   Measurement layout

antennas - the bright upper trace and the dim lower trace are
from the common antenna. The two other bright doppler traces
below the A scan are used to resolve directional ambiguities.
The *spikes* on these traces result from sawtooth phase modulation of the CW transmitter, and trace a *phantom wave* which
leads the normal doppler wave by approximately 90° if the trail
is drifting towards the receiving site, and lags the normal
doppler if the trail is drifting away from the receivers, thus
defining the sense of the doppler return. These films are read
on a semi-automated film reader, and one computer card is
punched for each echo, with date, time (to the nearest 15
minutes - if the orbit determination equipment is also operating, time on its film is to the nearest minute) echo number
(for correlation with the films from the meteor velocity and
wind shear displays), range, and the linear measurements from
the doppler and DF traces on the film required for reduction
to doppler frequency (with sign) and direction of arrival of
the echo. These cards are subsequently processed by a CDC
3600 computer to produce a tape for input to the subsequent
wind analysis program.

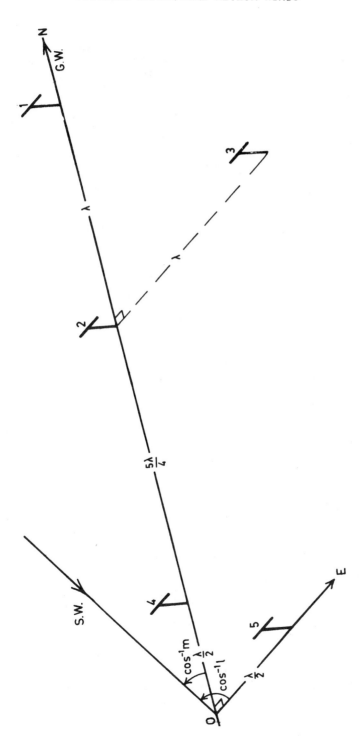

Fig. 8.3 Physical layout of the main receiving station antenna array.

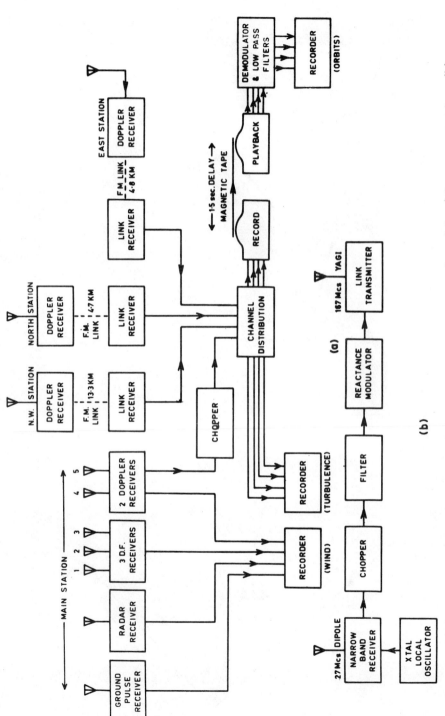

Fig. 8.4 Block diagram of equipment at the main receiving station (a) and an outstation (b).

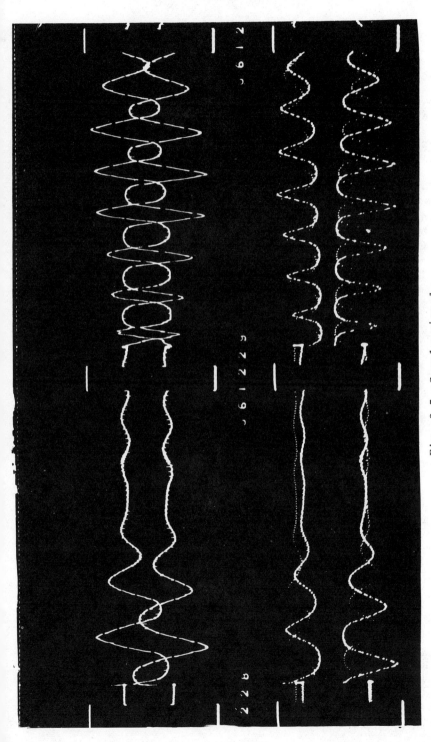

Fig. 8.5 Sample signal.

8.3 Data Analysis

Deduction of the *systematic features* of the actual wind
from the observed line of sight velocities is carried out
using a method of analysis developed by Groves (1959). In
this method it is assumed that, in general, the EW, NS, and
vertical components of the actual wind can each be expressed
as certain specified functions of height and time. The wind
is assumed to be *periodic* in time (in particular, to be made
up of the sum of a prevailing 24, 12, and 8 hour components)
and to have a *polynomial variation with height*.

The values of the amplitudes of the time dependent components, and the coefficients of the polynomials describing
the height variations, which give a wind model that best fits
the experimental data, are determined by the method of *least
squares*. In a typical analysis of data extending over some
ten days the number of coefficients may range from 70 to 100.
The standard deviation of each coefficient is also determined,
and a comparison of the magnitude of a coefficient with its
standard deviation is used to measure the significance of
that particular coefficient.

The model coefficients are combined to produce height-
time profiles of the mean zonal, meridional, and vertical
winds for a *typical* day of the month in question. The mean
wind is considered to be composed only of prevailing, 24, 12,
and 8 hour components.

For observations extending over intervals of time of the
order of one hour, the least squares analysis is carried out
by assuming no variation in time.

The fundamental periodicity assumed need not be 24 hours
as quoted above; the analysis is completely general, and any
fundamental period can be chosen. Thus, it is possible to
carry out a spectral (or, perhaps more precisely, a periodo-
gram) analysis of the wind data. Some examples of this type
of analysis are given by Roper (1966c) and Elford (1968), and
are contained in subsequent presentations here.

To illustrate the data reduction techniques, the analysis
for the month of September, 1961 is presented in Fig. 8.6.
Ten days data have been analyzed, using the previously mentioned model matching technique of Groves. The ten days data
are Fourier analyzed to determine the magnitude of the mean
(prevailing) wind, and the magnitudes and phases of the 24, 12
and 8 hour components. These are combined to produce the
height/time plots, for a *typical* day, of Fig. 8.6. The con-

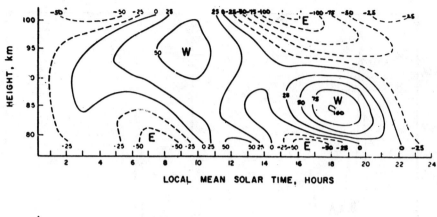

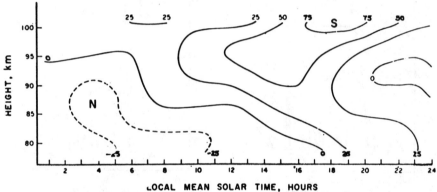

Fig. 8.6 Height-time profiles of upper atmosphere winds determined by the radio meteor technique at Adelaide, 35°S, in
September, 1961.

ventional meteorological terminology is depicted - E is an
easterly, a wind flowing from the east toward the west;
similarly, N is a northerly, blowing from the north towards
the south.

     The sums of the squares of the zonal and meridional wind
velocities for periods ranging from 0.5 to 4 cycles per day,
in increments of 0.05 cycle per day, are plotted from three
heights in Fig. 8.7. The significant feature, characteristic
of equinoxial months, is the dominance of the diurnal periodicity. That this periodic component is, in fact, the expected
diurnal tidal component is illustrated in Fig. 8.8, which
presents an hour by hour plot of the zonal diurnal component
over the 80 to 100 km height range. The results show that
there is considerable variability in the diurnal component

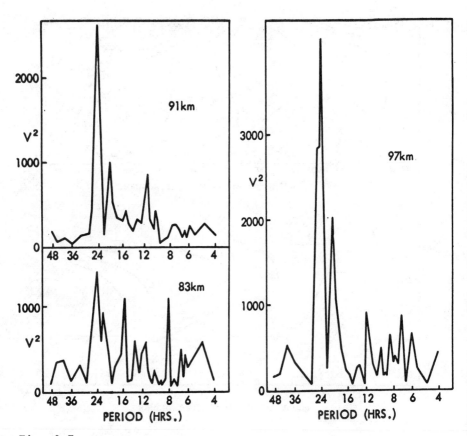

Fig. 8.7 Periodograms of the wind energy per unit mass for the month of September, 1961.

throughout the day. The divisions into *times* presented in Fig. 8.3, being centered on 24, 6, and 18 hours are somewhat arbitrary, but do illustrate that the wave structure is certainly better established at dawn and dusk than at midnight (1 a.m.?) or midday (13 hrs.?). An increase in wavelength with height is evident, together with an overall decrease in wavelength, as the wave becomes better established around dawn and dusk. Downward propagation of phase is evident at the higher altitudes, consistent with an upward propagation of tidal energy. The *90 km wavelength* at dawn and dusk is about 24 km, consistent with a slightly damped (1, 1) mode, and considerably greater (with much smaller amplitude) at midnight and midday, consistent with either greater damping of the (1, 1) mode, or interaction with other diurnal tidal modes (see Chapter 4).

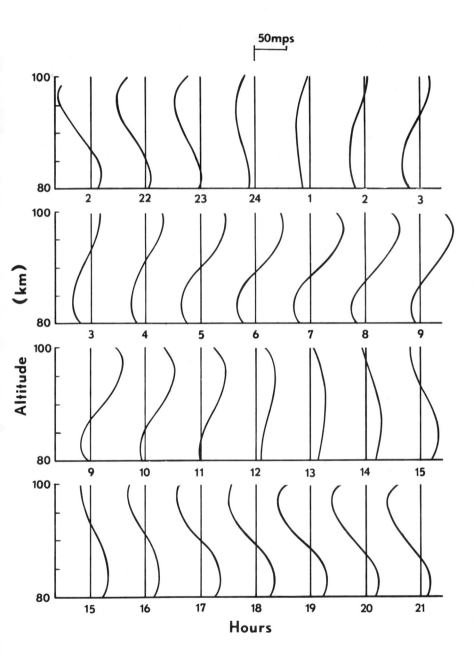

Fig. 8.8   Height variation of the diurnal tide observed in September 1961.

Having looked at one month's results in some detail, we will now turn our attention to the overall picture of the mean atmospheric dynamics deduced from the Adelaide measurements, bearing in mind that we do not, as yet, know a great deal about the latitudinal (and perhaps even the longitudinal) variations in the wind field at meteor heights.

8.4 Prevailing Winds

Over the height range 80 to 100 km the zonal wind is predominantly *toward the east,* with a tendency for a wind reversal to occur in the upper part of the height range in winter (Fig. 8.9). No significant reversals occur regularly

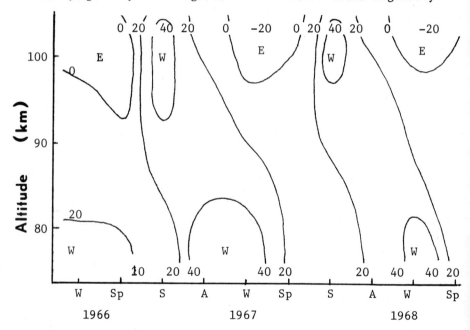

Fig. 8.9 Zonal wind component at Adelaide (35°S).

at other levels, but the wind has its minimum eastward amplitude in the spring at 90 km, and in the early summer at 80 km. As a result of the rapid change in wind with height, the seasonal patterns at 80 and 100 km are almost opposite in phase. This behavior is also reflected in the wind gradients, which have maximum values of +4m/sec/km in summer and -4m/sec/km in winter.

In contrast to the zonal winds, the meridional winds (Fig. 8.10) exhibit a behavior which is similar at all levels.

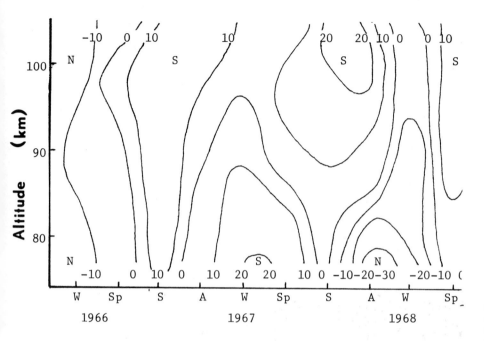

Fig. 8.10 Meridional component at Adelaide (35°S).

In general, northward winds occur during summer and southward winds during the winter. Based on the meridional variations measured at Jodrell Bank (53°N), College, Alaska (65°N), and Mawson (67°S), the meridional flow at these levels is consistent with a *horizontal movement of air from the summer to the winter pole*. The Adelaide results show that the mean meridional wind increases with height over the range 80 to 100 km, and that above 90 km the amplitude of the mean meridional wind is comparable to that of the mean zonal wind.

8.5 Diurnal Variations

A detailed investigation of the phases of the EW and NS diurnal components for each month indicates that the wind vectors rotate *counterclockwise* as is required for tidal motions in the southern hemisphere. The main features of the annual behavior of these components are best illustrated by grouping the months into seasons and determining the mean value for each season. The phase of the 24 hour component advances with height, particularly at the equinoxes (when, in general, the diurnal component has its maximum amplitude, typically 40 mps; see Fig. 8.11). According to tidal theory, the amplitude of the wind oscillation should increase with height

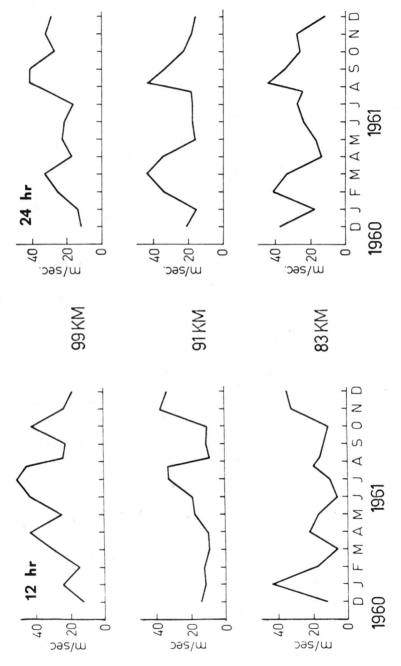

Fig. 8.11 The seasonal variation of the semidiurnal and diurnal wind components as measured at Adelaide during 1961.

such that

$$\rho u^2 = \text{constant}$$

where ρ is the ambient density, and u is the amplitude of the
tidal mode under consideration. The amplitude of the diurnal
oscillation as measured at Adelaide is approximately constant
over the height range 80 to 100 km, indicating severe damping
and energy dissipation (Roper, 1966b).

8.6 Semidiurnal Variations

The most marked feature of the mean seasonal values of
the semidiurnal component illustrated in Fig. 8.11 is the
reversal in phase from summer to winter, with maximum amplitudes (again, typically 40 mps at 90 km) occuring at the
solstices. The magnitude of the semidiurnal component in
winter and spring shows a strong positive height gradient
(almost no damping), but the phase shift with height is
relatively small in all seasons.

8.7 Terdiurnal Variations

The magnitudes of the 24- and 12-hour periodic components
have always been sufficiently large that there has never been
any question of their validity in the Adelaide results. The
possibility of there being a significant 8-hour wind component
had never been discounted, but the least squared analysis
showed that in general its magnitude was only marginally
greater than the root mean square deviation. In order to test
the reality of this component and to look for any other significant periodic components, each month of the set of observations made in 1961 was subjected to a frequency analysis.
This was done by using the previously mentioned generalization
of Groves' (1959) analysis to determine the amplitudes and
phases of all periodic components in the range 0.5 cycle/day
to 4 cycles/day in increments of 0.05 cycle/day.

To assist in identifying significant peaks in the spectra,
the wind energy per unit mass was calculated for each frequency
by forming the sum of the squares of the zonal, meridional,
and vertical components of that particular frequency. These
results have been reported by Elford (1968). In individual
months, there was a suggestion of a peak in the spectrum at
8 hours. To investigate the significance of the 8-hour wind
component through the complete year, the energy spectra for
the twelve months of 1961 were averaged (Fig. 8.12). The
24 and 12 hours components are dominant, as expected, but the
8 hour peak remains, while noise peaks comparable to the 8-

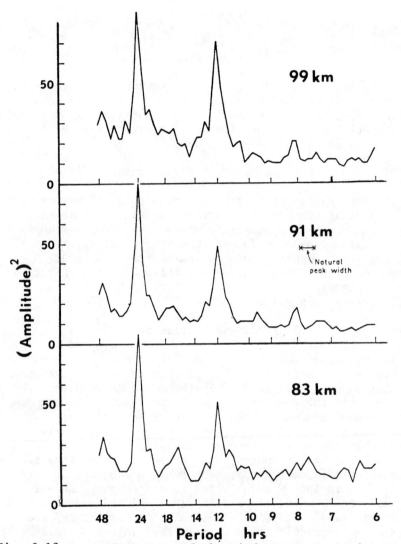

Fig. 8.12 Average spectra of the wind energy per unit mass at three heights for the period January to December 1961.

hour component in individual months have disappeared. It is thus considered that the terdiurnal component is a real feature of the winds in the 80 to 100 km region.

8.8 Wind Variability

Until recently, the meteor echo rate obtained with the Adelaide system was, in general, insufficient to enable the main features of the wind to be determined from a single days' observations. In principle, the continuous behavior of the wind during a given observing interval can be obtained by a

Fourier transform of the frequency spectrum of the amplitudes
(with phases) of the zonal, meridional and vertical components.
The extent to which this transformation gives a meaningful
result depends on the meteor echo rate in relation to the
highest frequency in the spectrum. Thus the deduced wind
patterns will be least reliable at times of low echo rate.
The striking feature of the results of this procedure is the
variability of the wind pattern from day to day at all levels.
This must be contrasted with the year to year repeatability
of the *typical* day's winds each month resulting from the
lumping together of from 5 to 10 day's data.

The extent to which the day to day variability is due
to large scale *weather systems*, gravity waves or turbulence
is not known. The current program of hour by hour profile
determination at Adelaide should lead to a more detailed
understanding of wind variability at these altitudes.

8.9 Turbulence

In addition to the mean and periodic motions detailed
above, observations made from a single station can give information about the characteristics of the *large scale random
motions*. The wind vectors as measured by the meteor method
show apparent random fluctuations; however, it is found that
the distribution of the velocity fluctuations indicates a pre-
ferred value (Fig. 8.13). The most probable value of the

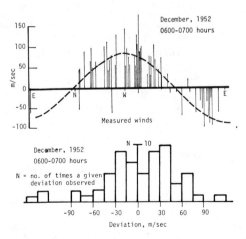

Fig. 8.13 Horizontal projections of the observed wind vectors,
the least-squares-fit vector (dashed curve) corresponding to a
mean wind of 86 m/sec directed toward the west, and the deviations from the mean for the hour 0600 to 0700, December 1952.

velocity characteristic of the large scale fluctuations
measured month by month at Adelaide has been found to range
from 25 mps to 40 mps; during individual months, characteristic velocities as low as 15 mps and as high as 60 mps have
been measured. It should be noted that, in general, the
irregular component of the wind velocity is less than half
the mean wind (as given by the sum of the prevailing, 24-,
12-, and 8 hour components), which means that some 25% of
the wind energy is in the irregular component.

Instantaneous shear on individual trails can also be
measured from a single receiving site by measuring the rate
of change of doppler beat with time as the trail is sheared
by wind gradients (Roper, 1966a). This method of shear determination has limited usefulness, in that only shear normal
to the trail is measured, and the orientation of the trail in
space cannot be determined from a single station observation.

Since radio reflection from meteor trails is specular,
the spacing of receivers on the ground makes it possible to
examine separate parts of the trail simultaneously. From a
knowledge of the echo range and arrival angle, as measured
at one station, and the meteor velocity and its time of
arrival at each specular reflection point, as measured from
the Fresnel diffraction pattern buildup at each of at least
three receiving stations, not only the distances separating
the reflection points, but also the orientation of the trail
in space, can be calculated.

The Adelaide three station system was operated from
December, 1960 to December, 1961 inclusive, and has been
described by Weiss and Elford (1963). Because of the limitations imposed by station geometry, and the subsequent behavior
of trails after formation, this three station survey produced
shear spectra over a range of reflection point separations
from 0.5 to 3.5 km. Turbulence theory shows that, for an
isotropic and inertial region, the mean square velocity difference measured between points separated by a distance r,
usually termed the structure function, should vary as $r^{2/3}$,
and that, if such a variation holds, the constant of propor-
tionality gives a measure of the rate of dissipation of turbulent energy in the region.

Application of the structure function method of analysis
to multistation meteor drift observations leads to two different results, depending on whether the separation is taken
in the vertical or in the horizontal directions. In the
case of horizontal separations, the exponent is approximately
0.7, a value that is quite close to that predicted by Theory

(Fig. 8.14). The more rapid variation of the structure function with vertical separation (Fig. 8.15) has also been observed in the case of chemical trails laid through this region

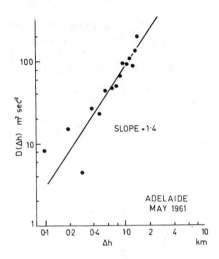

Fig. 8.14 Variation with height difference Δh of the structure
function D(Δh) determined from radio meteor trail shears at
Adelaide (35°S).

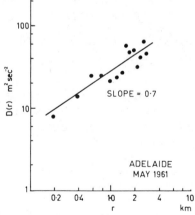

Fig. 8.15 Variation of the spatial structure determined from
radio meteor observations at Adelaide.

by rockets. In view of the inhibition of vertical motion by
buoyancy forces (Elford and Roper, 1967, Justus and Roper,
1968), it is probable that the height dependence of the
structure function is related to the gravity wave spectrum at
these levels, rather than to turbulence. A more detailed

examination of the meteor drift observations shows that in the
height range accessible to measurement, 80 to 100 km, the
turbulence is markedly anistropic for scales larger than a
kilometer.

Values of the rate of dissipation of turbulent energy
has been estimated for each month of 1961 using the spatial
structure function. As can be seen from Fig. 8.16, there is
a pronounced seasonal variation in the rate of dissipation of

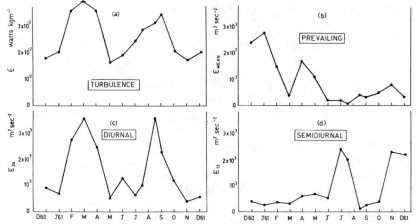

Fig. 8.16   Seasonal variations of turbulence dissipation, and
prevailing wind, diurnal and semidiurnal energy at Adelaide.

turbulent energy, minimum values or 1.5 x 10^{-2} watts/kgm
occurring in the summer and winter, with maxima of 3.5 x 10^{-2}
watts/kgm at the equinoxes (Roper and Elford, 1963)). Since,
as has been shown by Roper (1966b), the rate of dissipation
of turbulent energy increases rapidly with height up to approximately 105 km (Fig. 8.17), it is important to note that the
dissipation rates quoted here refer to the mean echo height
of 93 km.

It is interesting to note that extrapolation of the
measured shears *horizontally* to the velocity characteristic
of the large scale random fluctuations mentioned earlier
yields a horizontal scale of from 50 to 250 km, while a similar extrapolation *vertically* yields a vertical scale of from
5 to 9 km. These dimensions, together with the range of
characteristic irregular wind velocities measured, have been
identified by Hines (1960) as the dominant mode of the internal atmospheric gravity wave spectrum at these altitudes. Thus
the turbulence has as its energy source, random internal atmospheric gravity waves. As is also shown in Fig. 8.16, the
seasonal variation in turbulence is extremely well correlated

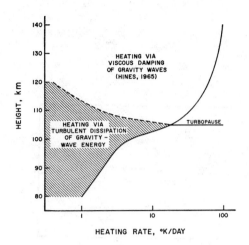

Fig. 8.17 Dissipation of gravity wave energy.

with the diurnal tidal wind, which is known to undergo severe
dissipation at these heights (Hines, 1966 Roper 1966b)
Figure 8.18 illustrates well the differential damping of the
diurnal oscillation. The diurnal tidal wind energy per unit
mass increases by a factor of only 1.4 from 83 to 97 kilometers, while the semidiurnal component increases almost exponentially, characteristic of an undamped propagating mode.
Thus the dissipation of the diurnal tide probably proceeds as

diurnal tide → gravity waves → turbulence → heating.

The 1961 seasonal survey is being repeated by the University of Adelaide this year, this time using five receiving
stations, located such that reflection point separations of
up to 20 km can be measured. Since a considerably higher
echo rate has now been achieved, diurnal as well as seasonal
variations, and possibly variations with height also may be
determined.

8.10   References

Elford, W. G., 1968. "Upper atmospheric wind observations at
Adelaide." Proceedings of the Workshop on Methods of Obtaining Winds and Densities from Radar Meteor Trail Returns, edited
by A. Barnes and J. J. Pazniokas, Air Force Cambridge Research
Laboratories, 68-0228, p. 195.

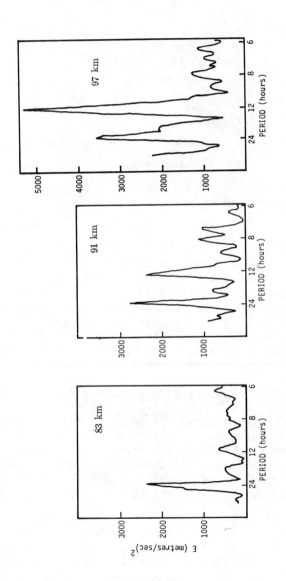

Fig. 8.18 The energy spectrum of meteor winds measured at Adelaide (35°S) in June 1961. The ordinate E is the sum of the squares of the north-south and east-west components.

Elford, W. G., and R. G. Roper, 1967. "Turbulence in the Lower Thermosphere." SPACE RESEARCH VII, North Holland Publishing Company, Amsterdam, p. 42.

Groves, G. V., 1959. "A theory for determining upper atmosphere winds from radio observations on meteor trails." Journal of Atmospheric and Terrestrial Physics, 16, 344.

Hines, C. O., 1960. "Internal Atmospheric Gravity Waves at Ionospheric Heights." Canadian Journal of Physics, 38, 1441.

Hines, C. O., 1966. "Diurnal Tide in the Upper Atmosphere." Journal of Geophysical Research, 71, 1453.

Justus, C. G., and R. G. Roper, 1968. "Some Observations of Turbulence in the 80 to 110 km region of the Upper Atmosphere." American Meteorological Society, Meteorological Monograph, 9, (31), 122.

Robertson, D. S., D. T. Liddy and W. G. Elford, 1953. 'Measurements of winds in the upper atmosphere by means of drifting meteor trails." I. Journal Atmospheric and Terrestrial Physics, 4, 255.

Roper, R. G., 1966a. "Atmospheric Turbulence in the Meteor Region." Journal of Geophysical Research, 71, 5785.

Roper, R. G., 1966b. 'Dissipation of Wind Energy in the Height Range 80 to 140 kilometers." Journal of Geophysical Research, 71, 4427.

Roper, R. G., 1966c. 'The semidiurnal tide in the lower thermosphere." Journal of Geophysical Research, 71, 5746.

Roper, R. G., and W. G. Elford, 1963. 'Seasonal Variations of Turbulence in the Upper Atmosphere." Nature, 197, 963.

Weiss, A. A., and W. G. Elford, 1963. 'An equipment for combined geophysical and astronomical measurements of meteors!' Proceedings of the Institute of Radio Engineers, Australia, 24, 197.

CHAPTER 9

RADAR OBSERVATIONS OF METEOR WINDS
ABOVE ILLINOIS

M. D. Grossi
R. B. Southworth
S. K. Rosenthal

Smithsonian Institution,
Astrophysical Observatory,
Cambridge, Massachusetts

9.1 Introduction

The Smithsonian Astrophysical Observatory (SAO) has developed a method for measuring *winds* from radar echoes of
meteor trails that stands apart from other Meteor Trail Radar
(MTR) approaches. The SAO method (Southworth, 1966) is based
on measurements of astronomical parameters pertaining both to
the electron trail left behind by the meteoroid in entering
the earth's atmosphere and to the motion of the meteoroid
itself. The method is capable of measuring the *horizontal and
vertical components* of the wind pattern in a specific volume
of the atmosphere (height range 80 to 100 km) without resort-
ing to horizon-to-horizon spatial averaging and with a sampling
rate sufficient to describe adequately the expected time changes
of the pattern. It has the height-measuring accuracy of about
1 km that is necessary to obtain meaningful correlations between instantaneously measured wind parameters and to construct
plausible average patterns. The height measurement is based
on the processing of geometrical quantities and does not rely
on diffusion-rate determinations.

The SAO meteor radar network (Deegan et al., 1970) is a
multistatic phase-coherent system that records on magnetic
tape and punched paper information regarding meteor echoes.
The network is located near Havana, Illinois, and comprises a
main site, five outlying stations, and two remote stations.
The outlying stations are located between 7 and 30 miles from
the main site, with microwave links providing for synchroni-
zation and transmission of data to the central point. The
coordinates of the stations are given in Table 9.1, along with
their locations relative to the main station.

Table 9.1   Network coordinates

Station number	Latitude	Longitude	Bearing East of North from site 3	Distance from site 3
1	40°19'34"	90°15'22"	300°44'53"	14.39 st.mi
2	40°17'36"	90°08'08"	310°08'57"	7.85
3	40°13'11"	90°01'19"	-------	------
4	40°18'44"	89°57'46"	26°09'16"	7.09
5	40°09'47"	89°54'32"	123°11'36"	7.17
6	40°09'23"	89°45'45"	107°34'17"	14.41
7	40°25'02"	89°31'44"	62°14'20"	29.39
8	39°54'36"	90°00'51"	178°53'03"	21.39

The network is a rework of the six-station noncoherent pulse radar system operated in Illinois since 1961 by the Harvard Meteor Project.

The radar system is able to collect phase-coherent meteor echoes from a volume in the upper atmosphere that is not sharply bounded but that measures roughly 50 km x 50 km horizontally and 20 km vertically, with the vertical dimension extending from about 80 to 100 km above the ground. The volume under radar patrol is approximately above Decatur, Illinois.

For analysis of the spatial structure of the wind field, we divide the observable volume into *cells*. We take 1 km as the vertical extent of a cell, approximately our accuracy of height measurement. Subject to later revision, we take 15 km as the extent of a cell in both horizontal directions; this is the order of distance in which we expect to be able to resolve differences in horizontal-wind components. Accordingly, we observe on the order of 220 wind cells.

When a cell is crossed by a meteor trail, echoes are received at the main site if the trail in the cell is tangent to a sphere that is centered at the main site and has a radius equal to the distance of the cell from this site (the back-scattering case). An outlying or a remote site then receives an echo when the trail is tangent to the ellipsoid having as foci the receiving station under consideration and the transmitter location (main site).

At the main site (where a 4-Mw peak power VHF transmitter is located), a dual-channel receiver provides range, angle of arrival, Fresnel pattern, and doppler information (radial velocity) on the echoes backscattered by a trail.

At each one of the outlying and remote sites, a single-channel receiver provides, for the echoes scattered to the site by the trail, information concerning range, Fresnel pattern, and doppler shift. The doppler information provides the wind-velocity component perpendicular to the ellipsoid determined by the echo circulation time from the main site to the meteor to the outlying (or remote) site.

A three-dimensional wind determination in a certain cell (measurement of the wind vector in its horizontal and vertical components) requires that echoes from the cell be received simultaneously at the main site (or, equivalently, at one of the outlying sites) and at both remote sites (or, for a less accurate measurement, at one remote site and two outlying sites distant from each other). Because, in general, a single trail does not scatter echoes in all the needed directions, we have to wait until the cell is crossed by more than one trail nearly at the same time in order to obtain the basic echoes required for three-dimensional wind measurements.

Assuming (conservatively) that 15 samples per hour for each cell are necessary for an adequate description of the temporal variations of the wind, 3333 independent samples per hour are required in the overall volume surveyed.

Operation of the Havana system since its completion has shown that the average hourly rate of reducible meteors, from after midnight to midafternoon, is approximately 250/hr. This means that we are able to collect 2000 independent samples per hour (every meteor trail provides an average of four segments, and each segment gives two pieces of information - wind velocity and its derivative) and to make about 1000 one-dimensional velocity measurements per hour, 40 two-dimensional measurements per hour, and 1 three-dimensional wind-velocity measurement per hour.

From midafternoon to shortly after midnight, the average rate of reducible meteors is of the order of 50/hr. In this period, we take the horizontal extent of a wind cell to be the same as the observable volume. The resolution in wind measurement which may be obtained is thus variable diurnally. The basic wind-measurement recording and analysis procedure is outlined in the following sections.

The measurements are mechanized by conveying to the main site (for processing in a digitizer and recording in a multi-channel digital tape recorder) all the data collected by the eight stations and by playing back the tape in a CDC 6400

computer appropriately programmed to print out the wind profiles. The computer program reads the tape, where 34 channels
of information are recorded, processes the data to obtain
positional information of the meteor trail and of its motion
due to the wind, smoothes these data and prints out the results. These data may then be placed in the desired format
for application to meteorological problems.

9.2 Considerations on Monostatic Versus Multistatic Radar Systems

     In general, the determination of the instantaneous velo-
city vector and of the position in three-dimensional space of
a target by radars having low-directivity antennas requires
the use of at least three monostatic pulse-doppler radar stations (each equipped with transmitter and receiver) located
at the vertices of a triangle with sides preferably comparable
in length with the distance of the target from each one of
the vertices. In other words, a point target in this configuration is at the apex of a pyramid that has a triangular base,
the other vertices (of the base) being the three radar stations.
A line target, such as a meteor trail, lies tangent to three
spheres centered at the stations. If the scattering properties
of the target are such that sufficient energy is scattered
toward the ground in forwardscatter as well as in direct back-
scatter, the system can be simplified to a multistatic configuration with a single radar station (transmitter and receiver) at one of the vertices (main site) and only receivers
at the other two vertices (remote sites). The main site
radiates the RF pulses and receives backscattered echoes; the
remote sites receive forward scattering.

     As far as target-motion determination is concerned, there
is this minor difference between the two solutions: with the
three monostatic radars, each station measures the velocity
component along the line of sight from the radar to the target.
In the *multistatic* scheme, this applies only to the main site,
while, for the two remote sites, the velocity component is
measured in the propagation plane along the bisector of the
angle that separates the main site and remote site as seen
from the target.

     As far as ranging is concerned, the determination of the
location in three-dimensional space of the target is obtained
(when three monostatic radars are used) from the intersection
of three spheres, each centered at one of the stations and
each with a radius equal to the delay between the target and
the station.

In the multistatic case, two of the three spheres become ellipsoids, each with a focus at the main site and the other at each one of the remote sites. An ellipsoid is identified by the measured circulation time delay of the echo in the propagation path from the main site to the target to the remote site. The target (meteor trail) location is found as a line tangent to the two ellipsoids and the sphere. There may be an ambiguity of solution; more important, the available accuracy in range measures may not yield a good position determination. To find the target position better, extra information can then be obtained by using interferometric arrangements at one or more of the sites and measuring echo angles of arrival. For each interferometer, a cone is identified at every echo reception, which provides additional intersecting surfaces for target positioning.

9.3 Scattering Properties of a Meteor Trail

A simplified picture of the scattering phenomenon can be obtained by considering that a radiowave at VHF (as in our case) incident on the column is scattered by individual free electrons, each of which oscillates as if no other were present (underdense trail condition). The *specular reflection point* (for a station or pair of stations) is the point on the trail at minimum distance from the station, or minimum sum of distances from the pair of stations. Echoes from electrons near the specular reflection point arrive in phase and are observed; echoes from points distant from the specular reflection point arrive out of phase and cancel each other.

Of the whole trail, the effective length that contributes to the scattering is the first *Fresnel zone* around the specular reflection point, half of which equals $\sqrt{R_0\lambda/2} \simeq 1$ km, in the normal incidence case, for monostatic radar (R_0 is the distance of the trail from the radar station, and λ the wavelength). In bistatic radar cases, this zone has a more complicated expression for oblique incidence. Different pairs of stations receive scattered signals from different parts of the trail.

Upper atmospheric *winds displace* the effective specular reflection points and physical processes alter the amplitude of received echoes, but both effects are almost always of second order in determining the position of a trail. They are discussed below.

The scattering properties of a meteor trail dictate the use of a multistatic system rather than widely spaced monostatic radars. If the monostatic radars are separated

adequately for good position determination (more than 50 km
on the ground), the specular reflection points are much far-
ther separated than the usual length of a meteor trail, so
that no meteor is observed at all three radars. On the other
hand, the separation of the specular reflection points is
directly useful in position determination with a multistatic
system.

9.4   Position of the Meteor Trail

The locations of the original six Havana stations, which
are the main transmitter and receiver site (site 3) and the
outlying receiver sites, were specially chosen for the determination of meteor-trail positions, in addition to other parameters. The locations of the two newer *remote* receiving sites
were chosen primarily for wind measures, but these sites also
assist in position determinations for meteor trails that they
observe.

We use the motion of the meteoroid that forms the meteor
trail (the *target*) to find the position of the trail. As the
trail is extended past a specular reflection point, the addition of electrons that are alternately in phase and out of
phase with the main echo from the principal Fresnel zone
generates a characteristic *Fresnel pattern* in the received
signal. From the spacing in time of the oscillations in this
pattern and the length of the principal Fresnel zone, we find
the velocity of the meteoroid (McKinley, 1961; Southworth,
1962). From that velocity and the spacing in time between the
Fresnel patterns observed at different stations, we find the
distances between specular reflection points. Two such distances, corresponding to noncolinear pairs of stations, determine the direction of motion of the meteoroid; additional distances may strengthen the determination.

At this stage in the reduction, we know that the specular
reflection point from the main site lies in a plane normal to
the direction of motion of the meteoroid, at an observed
range from the main site. (We also used the observed range
in computing the velocity.) The meteor trail lies on a cy-
linder normal to that plane. Measures of differential range
from our closely spaced sites are, in general, not sufficiently
accurate to fix the location of the meteor trail on the cylin-
der; but this location is determined by an interferometric
measure at the main site. Finally, a general iterative procedure uses all the observations in least-squares fits to find
the best position.

9.5 Wind Measurements

The echo received at each station conveys information concerning a part of the meteor trail (the principal Fresnel zone)
roughly 1 km long. In general, each station observes a different part of the trail, with some overlapping. The spacing
between centers of adjacent parts observed on the trail by
different stations is 0 to 3.5 km along the trail and 0 to
2.7 km in height.

Phase measurements at a station furnish the *radial* component of the mean wind velocity, averaged over the corresponding part of the trail. This component is directed along the
bisector of the angle between the directions to the transmitter and the receiver. We normally convert this to a *horizontal radial* component, on the assumption that there is no
vertical component; but the actual vertical component is sometimes computed, as described below. The phase measurements
also normally furnish the first derivative of the radial component with respect to position along the trail. Both the
wind and its derivative are also used to correct the determination of the meteoroid's position and direction of motion.

To determine two components of the wind velocity at a
point, it is necessary to receive an echo and measure its
phase at one of the remote sites and also to measure the
radial wind component (by phase measurements at one or more
of the stations of the cluster of six original stations) in the
same part of the trail. With phase measured at all the stations
of the cluster, the radial wind component can be found, by
interpolation, at every point of the observed portion of a
trail. Then every meteor observed at two or more stations of
the cluster, and also observed in the same portion of the
trail at a remote site, yields one two-dimensional wind determination.

For three-dimensional wind-velocity determinations, we
must note that virtually no trail can ever scatter an echo
toward the cluster and also toward both remote sites. Therefore, we have to use, for three-dimensional velocity determinations at a given point in space, two different trails that
cross the same point within a short interval of time, each of
which satisfies all the requirements for positional determination. Together, the two meteors must yield components from
one or both remote sites and two widely separated outlying
sites.

The detail possible in describing the wind field depends

upon the meteor influx rate, which is comparatively high from
shortly after midnight to midafternoon. During this part of
the day, the average rate of reducible meteors has been found
to be of the order of 250/hr; about 40 of these will also be
observed at one or the other of the remote sites. With all
stations equipped to measure phase, there are about 1000 one-
dimensional wind determinations and 40 two-dimensional determinations per hour. The number of three-dimensional determinations also depends on how close together the two meteors must
be in time and space. If 4 min, 15 km horizontally, and 1 km
vertically are allowed, then there is roughly one three-dimensional determination per hour.

The system can furnish vastly more information about the
wind than has been available heretofore. The amount of information necessary to specify the wind field if the pattern is
not known can be roughly estimated as follows. We take these
estimates for the volume of space surveyed and the scales of
the dominant variations in the wind:

Volume		Variation(wavelength)	
height	20 km	vertical	5 km
width	50 km	horizontal	75 km
length	50 km	time	20 km

We can assume (conservatively) that five points per wavelength
are required to describe a variation. Then the number of
independent pieces of information needed per hour is

$$\sim 5 \left(\frac{20}{5}\right) \times 5\left(\frac{50}{75}\right) \times 5\left(\frac{50}{75}\right) \times 5\left(\frac{60}{20}\right) = 3333.$$

In Havana, we routinely obtain one-half of this hourly
information. However, if we assume that four samples for
wavelength are sufficient for the descriptions of the variations, the information collected by the Havana system becomes
more than adequate.

From midafternoon to after midnight, the average rate of
reducible meteors is of the order of 50/hr. In this period,
we would subdivide the observing volume only into layers 1 km
thick but would look for linear horizontal variations of the
wind within each layer. If the dominant wavelength for horizontal variation is 200 km or more, as seems very possible,
the lower rate of meteors still yields a sufficient description
of the wind field.

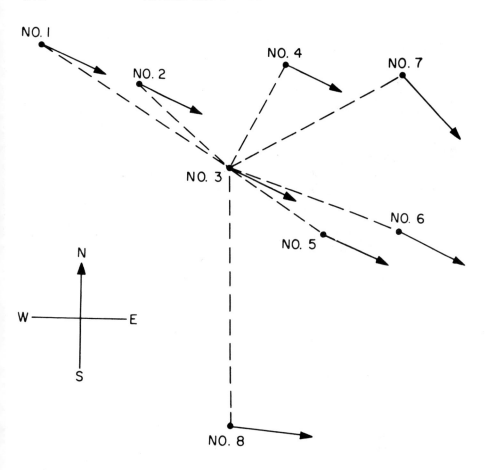

Fig. 9.1 Network layout. Dashed lines represent microwave
links. Arrows represent antenna axes: azimuth 113°East of
North at sites 1 to 6, 139° at site 7, and 94° at site 8.

9.6   The SAO Phase-Coherent Radar System

The basic scheme of the 40-92-MHz meteor radar built and
operated by SAO near Havana, Illinois, is depicted in Figs. 9.1
and 9.2.   The main site (No. 3) is the transmitter location,
with the receiver sites tied to the time base of the main site
through microwave data links. Data acquired by the receiving
stations are passed over these same communications circuits to
the main site.   In a phase-coherent network with stations
miles apart from one another, as in ours, the heart of the
problem is represented by the phase locking of the various

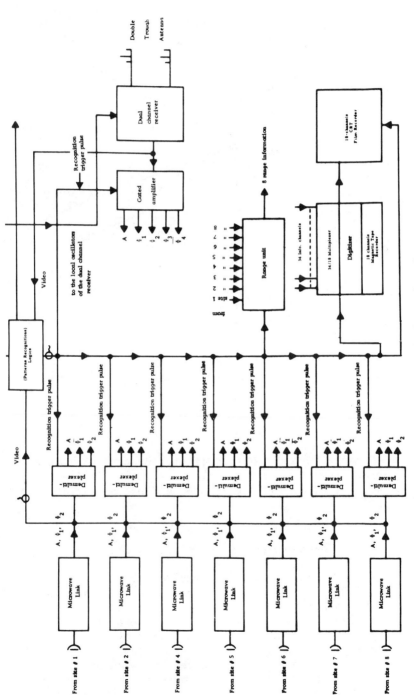

Fig. 9.2 Simplified block diagram of the main site.

sites among themselves. The requirement in our case is to
have a relative phase jitter between the transmitter and the
local oscillators of the various superheterodyne receivers of
the multistatic network no larger than 10% of the minimum
detectable doppler. For a signal-to-noise ratio of 20 db and
a meteor-trail duration of approximately 0.05 sec, the minimum detectable doppler corresponding to the radiated waveform
is 0.25 Hz (Grossi, 1963). Consequently, the relative phase
stability must be on the order of 5 parts in 10^{10}. The
Havana radar meets this requirement routinely by using a single
master oscillator at 1 MHz at the main site, from which, by
frequency synthesis, the frequency required for the transmitter
and the local-oscillator frequencies required by the local
dual-channel receiver have been obtained. Then, by dividing
down the 1 MHz to 2.5 kHz (400:1 ratio), a reference signal
is obtained that is distributed by telephone line to the
other seven sites of the network. This tone, however, cannot
be used directly at the remote end by multiplication and frequency synthesis because the telephone lines are not phase-
jitter free. By averaging phase observations for a minimum
interval of 10 min, we can obtain reliable phase reference.
At each outlying and remote site, the phase of the reference
tone is then compared with the received signal.

9.7 Outlying/Remote Station

Figure 9.3 shows a block diagram of an outlying/remote
site. In the calibration mode, the antenna is disconnected
from the receiver and simulated echoes of known amplitude
are sent to the receiver from the calibrator.

Control signals for the calibrator are received from the
master station via the microwave link in the form of switching
tones and the prf. The tones are received by a tone receiver,
which controls the calibrator.

The receiver produces three outputs. The amplitude (A)
pulse is proportional to the logarithm of the amplitude A_ℓ of
the echo. It is sent to the master station via the microwave
link on a 2.3 MHz subcarrier. The Phase 1 (ϕ_1) and Phase 2
(ϕ_2) pulses are proportional to $A_\ell \cos \phi$ and $A_\ell \sin\phi$, where
ϕ is the phase of the echo compared to the transmitted phase.
They are multiplexed in time and sent as video signals to the
master station via the microwave link.

An oscillator at the remote stations is phase-locked to
the phone-line reference to produce the basic frequency for the
synthesizer.

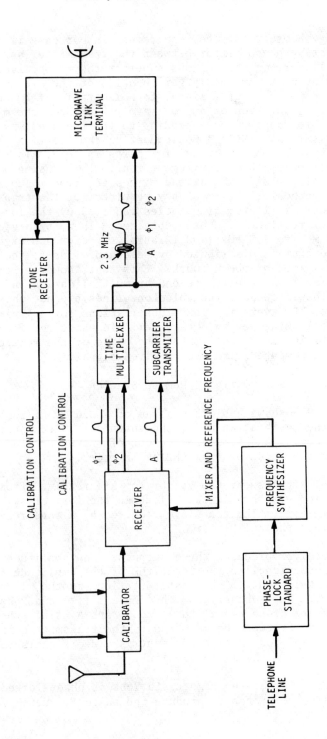

Fig. 9.3 Block diagram of the outlying/remote stations.

9.8 The Meteor Trail Radar

Figure 9.4 is a detailed block diagram of the master station. Only the blocks for one of the seven links to the outlying/remote stations are shown. The outlying/remote station data that come via the microwave link are sent to the low pass filter and subcarrier receiver. The time-multiplexed video signals ϕ_1 and ϕ_2 pass through the filter and the gain control to the demultiplexer. The subcarrier signal is blocked by the filter.

The subcarrier receiver and inverter amplifier reconstruct the A signal for the buffer delay gate. The buffer delay gate interfaces the echo recognizer and demultiplexer. The trigger for the demultiplexer is derived from the recognizer gate signal and the A signal. When the demultiplexer receives the trigger, it gates out the selected A, ϕ_1, and ϕ_2 signals on separate lines for recording. The ϕ adder allows both ϕ_1 and ϕ_2 to be recorded on the same cathode-ray tube.

The processing of the data collected by the master station (main site) is slightly different because there is no need to multiplex the receiver outputs and because the two antennas are used in conjunction with the transmitter. The TR-ATR networks allow the same antennas to be used for transmission and reception. The master-station receiver's outputs A, ϕ_1, and ϕ_2 are the same as the ones of a remote/outlying station and are derived from one of the antenna inputs. The outputs ϕ_3 and ϕ_4 are the phase signals from the second antenna input of the double trough. The adding amplifiers are similar to the demultiplexer and gate the desired signals to the recording equipment. Two triggers are used, one for the amplitude A and one for the phases ϕ. The ϕ signals are time-multiplexed for the CRT recorder so that they can be distinguished on the CRT.

The calibrator control simultaneously operates the master calibrator directly and the remote/outlying calibrators through the tone transmitter and microwave links. The entire meteor trail radar system then operates as a unit, with all components under direct control of the main station. Problems associated with the large geographical extent of the system are thus minimized, and the operation can be controlled from a central point.

The Continental Electronics Manufacturing Company Type PO-830 transmitter used by the radar has been designed for operation at a single frequency in the range of 40 to 42 MHz with the following operating characteristics:

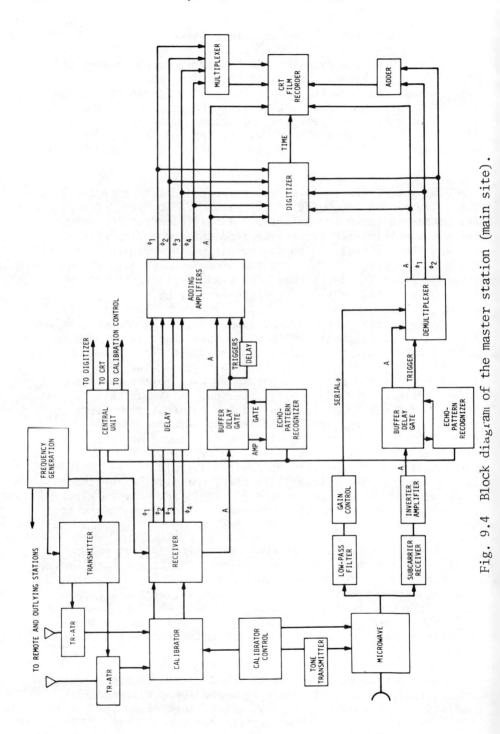

Fig. 9.4 Block diagram of the master station (main site).

Peak power output	4 Mw
Average power output	20 kw
Pulse waveform	3 µsec at 1330 pps to 100 µsec at 40 pps
Nominal duty factor	0.004
Power supply requirements	230 v, 3 phase, 60 cps
Power source capacity	200 kv-a
Type of emission	Pulsed CW
Output impedance	100 ohms balanced (two 50-ohm unbalanced outputs 180° out of phase)

Although the Continental PO-830 transmitter has a peak power-output capability of 4 Mw, it is normally operated at approximately 2 Mw. The transmitter receives the r.f. carrier input from the frequency synthesizer, while the modulating prf pulse input is generated within the meteor-radar processor and has a period of 1355 µsec. The quasi-gaussian radio frequency pulse has a 6-µsec half-power width. Range ambiguities up to 5 prf periods are resolved by a method of pulse skewing, in the meteor-radar processor, which delays every 5th pulse by 8 µsec.

At the main site, the output combiners of the transmitter's power amplifier transfer power through a TR-ATR network into a trough-guide antenna. This network automatically switches the antenna from the receivers to the transmitter during pulse transmission.

The TR-ATR system is a standard circuit adapted for co-axial line. The TR and ATR sections, which are spaced an electrical quarter-wavelength apart, each consist of a half-wavelength coaxial cavity with a tuning condenser and TR tube. The networks are fabricated from standard 9-inch-diameter co-axial line. The entire system is pressure tight for reliability.

Two asymmetric troughguide antennas are located side by side as a symmetric, integrated unit. The apertures are in phase for transmitting. During reception, each trough guide is used separately through the dual TR-ATR system. The antenna is horizontally polarized in the main beam and has a gain of 22 db when the apertures are connected in phase (feeds connected 180° out of phase).

The double trough-guide antenna measures 60 m in length, 21 m in width, and 10.5 m in height. The main beam is directed at an azimuth 113° East of North and has an elevation angle 35° above the horizon. The observed ionospheric volume measures roughly 20 km x 50 km x 50 km and has a characteristic distance from the transmitter of 150 km. The mean height of the volume above sea level is 90 km.

The five neighboring (outlying) stations (sites 1, 2, 4, 5, and 6) each have double trough-guide antennas connected with coaxial matching transformers to provide an input for a single-channel receiver. The two remote stations (sites 7 and 8) have 13-element Yagi antennas with a front-to-back ratio of 30 db, between 40 and 42 MHz. The outlying- and remote-station antennas are positioned to receive reflections from the same volume as the master station.

The frequency generation subsystem generates and distributes high-stability reference signals that cause the eight-station network to become phase-coherent. The frequency-generation subsystem is implemented with the master oscillator, the 2.5-kHz generator, the offset frequency synthesizer, and the telephone-line network (Fig. 9.5).

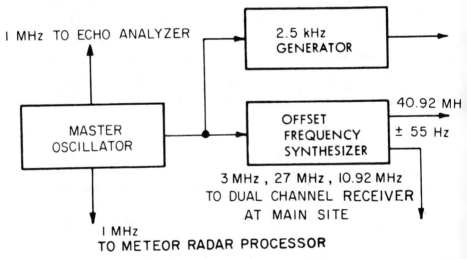

Fig. 9.5 Frequency generation at master station.

The master oscillator produces high-stability outputs of 1 MHz and 100 kHz. The 100 kHz output drives the 2.5 kHz generator, the output of which is provided to outlying and remote sites via telephone lines and to the offset frequency synthesizer and the transmitter. A Motorola oscillator, type S1054AL, is the basic system frequency standard (Fig. 9.6).

The output of the 1-MHz oscillator drives the meteor-radar processor, the echo analyzer, and part of the offset frequency

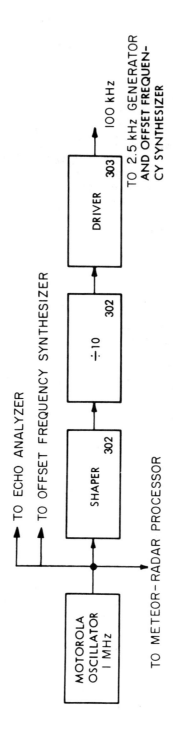

Fig. 9.6 Master oscillator.

synthesizer. Another output of the 1-MHz oscillator is divided
by 10 after passing through a shaper circuit and then enters
a driver circuit, which serves as the 100-kHz output of the
master oscillator.

An offset frequency synthesizer produces the transmitter
carrier frequency and has a switching means that allows the
basic 40.92 MHz frequency to be offset by 1-Hz increments up
to +55 Hz. This offset provision allows the system to simulate known doppler frequencies of 0 to +55 Hz in single-cycle
increments. Nonoffset frequencies of 3, 10.92, and 27 MHz
are generated by the synthesizer for the dual-channel receiver
present at the main site. The 100-kHz input frequency requires
that six multipliers and five dividers be used to generate the
offset units and tens digits. Two switches are used to select
the addition or subtraction operation. All other multipliers
and mixers in the synthesizer have frequency ratios of approximately 10:1 and reliably suppress spurious sidebands without
retuning. All frequencies to be rejected are at least 10%
away from center frequency, and all frequencies to be accepted
are at most 0.5% away from center frequency. The divider
circuits are fail-safe, producing no output when the input
falls below a minimum level.

A 2.5-kHz generator produces the continuous-wave synchronizing signal responsible for remote-site phase coherence.
The 100-kHz input signal passes through a divide-by-40 stage
consisting of one decade counter and two binary counters.
The resulting 2.5-kHz signal is fed through an emitter follower
driver stage, an amplifier, and a lightning protector and
then applied to a telephone hook point for distribution to the
remote sites.

The telephone lines carry the synchronizing signal from
the 2.5-kHz generator located at the main site to the phase-
locked secondary standards at all remote stations. This syn-
chronization allows the entire system to become phase coherent.
The frequency standards at all stations must be made phase
coherent in order that doppler frequencies of 0.3 to 300 Hz
can be accurately measured. The telephone lines are voice-
grade channels with a frequency response of 300 to 3000 Hz and
a level of at least -8 dbm into 600 ohms.

Each outlying/remote station contains an antenna, a single-
channel receiver, a phase-locked secondary standard, a frequency synthesizer, a multiplexer, a subcarrier transmitter,
a calibrator, a tone receiver, and a microwave link. The
2.5-kHz reference signal, brought from the master station to

the remote sites by telephone line, is connected to the phase-coherent 1-MHz and 100-kHz signals, which are operated on by the frequency synthesizer for the generation of the receiver-reference signals. The receiver describes target echoes by producing three parallel outputs: A, ϕ_1, ϕ_2. The signal A is sent on the subcarrier, while ϕ_1 and ϕ_2 are time-multiplexed and sent directly on the microwave, as bipolar video signals.

A block diagram of the phase-locked secondary standard is shown in Fig. 9.7. The phase-lock loop has a long time constant so that the short-term stability of the oscillator cannot be disturbed.

The 2.5-kHz synchronizing signal enters a lightning protection network that safeguards the input circuits of the secondary standard. The signal passes through a 2.5-kHz amplifier and filter that removes telephone-line noise. It then enters a phase detector, which performs a comparison between the reference signal and the 2.5-kHz driver output. The output of the phase detector is a 2.5-kpps pulse with its width proportional to the phase difference. The pulse feeds the integrator circuit for controlling the frequency of the 1-MHz voltage-controlled crystal oscillator. An isolator-driver stage prevents loading of the oscillator. A decade-counter divider reduces the frequency to 100 kHz. A second decade counter and two binary counters further reduce the frequency to 2.5 kHz for application to the phase detector, thereby closing the loop. The 2.5-kHz driver stage prevents loading of the divider circuits. The 100-kHz filter amplifier extracts the 100-kHz component from the first decade divider, providing a clean sine-wave output of adjustable amplitude for driving the remote-site frequency synthesizer.

The frequency synthesizer (Fig. 9.8) receives the phase-locked 1-MHz and 100-kHz sine waves from the secondary standard. Three phase-coherent signals are generated for the single-channel receiver. These signals consist of 10.92- and 27-MHz local oscillator frequencies and a 3-MHz reference phase-detector frequency.

The receiver is a one-channel version of the receiver used at the master station and produces amplitude, phase, and phase-sense information. The receiver is a double-conversion heterodyne unit followed by two phase detectors operating in quadrature. The 40.92-MHz echo returns are converted to 30 MHz, from which are derived both linear and logarithmic video data. The 30-MHz I.F. is also fed into a second mixer for conversion to 3 MHz, which is brought to the phase detectors.

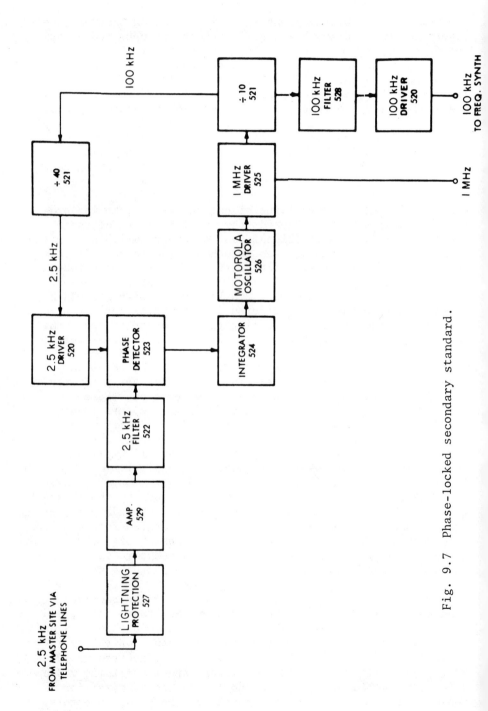

Fig. 9.7   Phase-locked secondary standard.

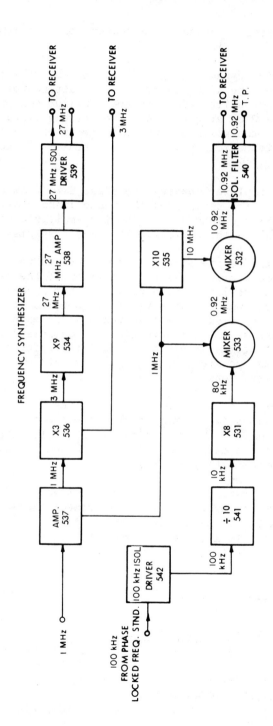

Fig. 9.8   Frequency synthesizer.

The phase detectors operate in quadrature to eliminate phase ambiguity.

The receiver dynamic range exceeds 60 db, of which 40 are available at any one time. A front panel attenuator provides selection of the desired 40 db range. Maximum sensitivity is obtained with a 0 db attenuation. When strong signals are being received, the maximum dynamic range is obtained by using the minimum attenuation that prevents linear I. F. amplifier saturation.

A time multiplexer obtains the phase data from the single-channel receiver and serializes the phase 1 and phase 2 information at 25-μsec intervals. The time-multiplexed signals are then brought to the microwave link for transmission.

The outlying/remote stations are connected to the main site by wideband microwave links that transmit the broadband amplitude and phase information to site 3 for processing. The microwave equipment transmits at frequencies of approximately 7 GHz. The signal-to-noise ratio is in excess of 50 db. Each of the seven microwave links is comprised of two terminals and contains a service channel for voice communication. All links operate over a single hop path without the need for relay stations.

The master station is connected to sites 1, 2, 4, 5, and 6 by MCR-1000 links, which have a r.f. power output of 1 w and a 2.3-MHz baseband. MW502A units, with a r.f. output of 100 mw, connect the main site with sites 7 and 8 and have a baseband of approximately 360 kHz.

The master station uses a 130-ft tower and a 200-ft tower. The 130-ft tower is used by the links between the master station and sites 2, 4, and 5. The 200-ft tower contains reflectors toward site 1 and the antenna toward site 6 at a 150-ft height, and reflectors toward sites 7 and 8 at the 2-0-ft height.

The purposes of the meteor-radar processor are the following:

A. To recognize, with an adjustable probablity of error, target echoes from interference, regardless of the amplitude of interference.
B. To determine which target's echoes and which interference signals are to be recorded.
C. To execute range measurements on the selected echoes.

D. To control automatically the motion of the tape and
film in order to have records of the proper length.
E. To track individual targets throughout their lives
and to maintain memory of them for a preselected duration after
their end in order to avoid repeated recordings of long-lasting
or fading echoes.
F. To generate auxiliary signals for the digitizer and
transmitter.

The meteor-radar processor is located at the master station (site 3). The processor is comprised of eight pattern
recognizers and one central unit arranged within two standard
19-inch relay racks. Each rack contains its own power supply
and circulating-air blower. Solid-state circuit cards, containing silicon semiconductors, are used throughout the units.

The pattern recognizers and central unit are connected
within the system as shown in Fig. 9.9.

The system has been organized into three distinct levels
of processing, i.e., recognition, decision, and control. This
organization results in a converging reduction of data.

The purpose of the first-level processing is to recognize target echoes and to track them throughout their entire
lives.

Each site has a corresponding pattern recognizer with the
capacity for recognizing 26 targets simultaneously. The eight
pattern recognizers may consequently perform first-level
processing on a maximum of 208 contemporary signals. In order
to execute this large amount of processing without a prohibi-
tive quantity of hardware, the first-level processing is performed with circulating self-programed digital words.

A delay medium for digital signals is connected in a
closed loop with an operating digital circuit. Digital bits
circulating within the system are grouped in independent words
that contain a 3-bit program part and a 13-bit data part. The
word capacity of the circulation system is determined by the
bit rate and the length of the delay line. When the word
enters the operating circuit, the program portion of the word
organizes the circuit for a particular operation to be performed on the data and program bits. After the operation is
executed, the modified word recirculates and once more enters
the operating circuit where an additional operation is executed.
In this way, a long process can be performed on stored or
arriving data, step by step, in many circulations. Moreover,

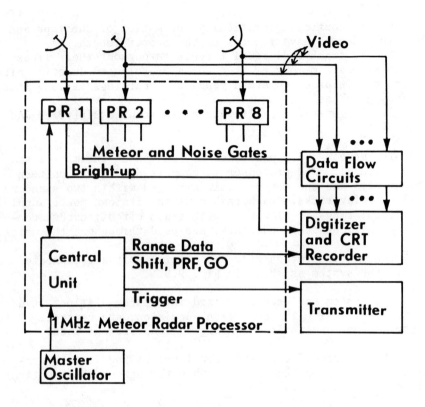

Fig. 9.9 Block diagram of meteor-radar processor.

many independent processings can be performed by several words
circulating simultaneously in the system.

The processing performed upon a word in circulation is
controlled by the program bits associated with that word.
These bits may have a configuration corresponding to the states
indicated in Fig. 9.10. Four states and six state transitions
are possible;

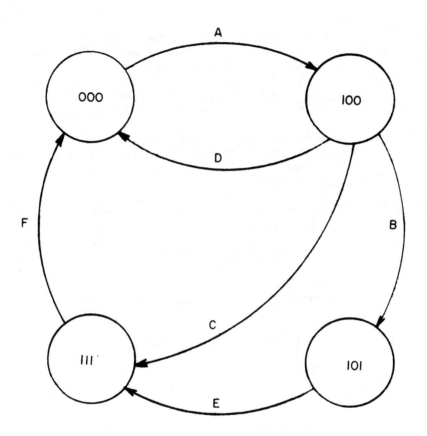

STATES	STATE TRANSITIONS
000	A
100	B
111	C
101	D
	E
	F

Fig. 9.10 States and state transitions.

000	The circuit is awaiting input data.
A	A detected signal causes a transition to state 100.
100	A word is placed in circulation. The presence or absence of a signal from the corresponding range is noted in succeeding circulations of the word, and the data bits are appropriately changed - a binary 1 is added or subtracted during each circulation.
B	The data count within the word reaches a preset value K of 4, 8, 16, or 32. A transition to state 101 occurs. With K = 16, the data bits are changed to 1000010010000.
C	The data count reaches the preset value, but another circulating word is currently in state 101. A transition to state 111 occurs.
D	The data count reduces to zero because of the absence of input signals from that range corresponding to this word. A transition to state 000 occurs.
101	A target echo recognition signal is generated. The word remains in circulation to track the target, but counting is restricted from 1000000000000 to 1000010011111.
E	A transition to state 111 occurs because the data count falls to 1000000000000 or because of an order from the central unit.
111	The word is kept in circulation with plus and minus counts continuing in order to track the target.
F	The data count reaches zero. A transition to state 000 occurs.

The counting operation permits the discrimination of unwanted signals from target echoes by utilizing arithmetical and logical functions.

Many targets are not stationary in space and consequently produce echoes with a time drift between sweeps. In order to track echoes that have motion, the circulation word period is allowed to undergo a self-correction in each circulation of $\pm$ 1 μsec, depending on whether the echo drift is advanced or delayed in time.

Decisions are made during the second-level processing that determine the data to be recorded. Each pattern recognizer allows only one of the many targets simultaneously recognized and tracked to be recorded at any single time. Recognized targets being tracked but not recorded may be subsequently recorded, but only if another pattern recognizer activates a record.

An additional function of this processing level is to select sample intervals for interference (terrestrial origin) and noise-signal (cosmic origin) recording.

Third-level processing is performed by the central unit after receiving signals from the second processing level. The purpose of this level is to control the recording devices for acceptance of data, thereby minimizing the use of magnetic tape and film.

This processing does limit the recording of long-duration targets to a predetermined maximum interval of their early life. A target existing longer than this interval may be recorded again but may not itself activate a new record. The range measurements are executed at this level, and the required auxiliary signals are generated.

The purpose of the Baird-Atomic digitizer is to accept 26 video input channels and one digital input and to record them digitally on half-inch magnetic tape with a format suitable for computer processing. The 27 input channels represent the amplitude outputs of all eight receivers, the ϕ_1, ϕ_2, ϕ_3, and ϕ_4 outputs of the master receiver, and a target-range channel from the meteor-radar processor. The digitizer recording is started and stopped by the central unit.

The digitizer (Fig. 9.11) consists of a 27-to-18 multi-plexer, an 8-bit analog-to-digital converter, logic circuits that provide both transverse and longitudinal parity, a time-of-day clock, and a magnetic-tape transport. The transport is a seven-track Potter MT 36 that incorporates vacuum buffering, 2-1/2-minute rewind, 15-sec tape threading, and 3-msec start time. This unit records at 36 inches/sec with a packing density of 556 bits/inch.

A playback feature provides an analog output of one of the 18 digital channels. The channel is selected by a panel switch on the digitizer.

9.9 Format of the Raw Data and Example of Digital Recording

The digital tape recorder has 18 channels, which record the following information for each pulse (i.e., each prf period) of the echo train, in one 8-bit word per channel:

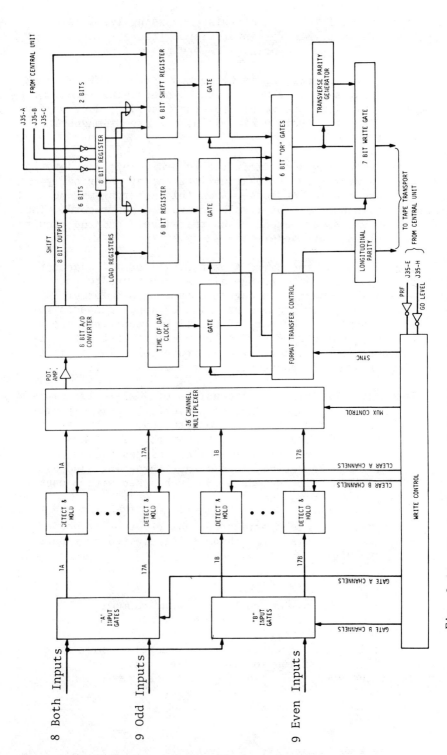

Fig. 9.11 Block diagram of digitizer circuits ("write-in" logics).

Channel 1 - Phase information (ϕ_1/ϕ_2) for station 1. The channel is time-shared between A sin ϕ_1, called $(\phi_1)_1$, and A cos ϕ_1, called $(\phi_2)_1$. Every pulse of the returned train is given a serial number. Pulses with odd numbers carry sin ϕ_1 information; pulses with even numbers carry cos ϕ_1 information. These are the two outputs of the phase detector. (A, the amplitude of the echo, is recorded on channel 2.) Amplitude calibration signals (11 levels) are injected into the system once per hour.

Channel 2 - Echo amplitude A_1 for station 1. Amplitudes are usually recorded on a logarithmic scale, unlike the linear scale used in the phase channels.

Channel 3 - Time-shared between $(\phi_1)_2$ and $(\phi_2)_2$, station 2, as in channel 1.

Channel 4 - Echo amplitude A_2, station 2.

Channel 5 - Time-shared between $(\phi_1)_4$ and $(\phi_2)_4$, station 4, as in channel 1.

Channel 6 - Echo amplitude A_4, station 4.

Channel 7 - Time-shared between $(\phi_1)_5$ and $(\phi_2)_5$, as in channel 1.

Channel 8 - Echo amplitude A_5, station 5.

Channel 9 - Time-shared between $(\phi_1)_6$ and $(\phi_2)_6$, as in channel 1.

Channel 10- Echo amplitude A_6, station 6.

Channel 11- Time-shared between $(\phi_1)_7$ and $(\phi_2)_7$, as in channel 1.

Channel 12- Echo amplitude A_7, station 7.

Channel 13- Time-shared between $(\phi_1)_8$ and $(\phi_2)_8$, as in channel 1.

Channel 14- Echo amplitude A_8, station 8.

Channel 15- Time-shared between $(\phi_1)_3$ and $(\phi_2)_3$, as in channel 1.

Channel 16- Echo amplitude A_3, station 3.

Channel 17- Time-shared between $(\phi_3)_3$ and $(\phi_4)_3$; phase information from the second receiver of station 3; time-shared as in channel 1.

Channel 18- Range information from that one of the eight stations of the network where the meteor was most recently recognized by the pattern-recognition logics. For every measure, 5 prf periods are used to provide adjacent time segments that are occupied by the range information. The first two carry the octal word 000 and are used as synchronizing information. The third prf period is used to identify the station according to the following correspondence table:

Station

200	1
100	2
040	3
020	4
010	5
004	6
002	7
001	8

The fourth prf period is used by an 8-bit word that gives the high-order digits of the range information (800-km maximum range). The fifth prf period is used by an 8-bit word that gives the least significant digits of the range information. The range information is the number of quarter microseconds between transmission and reception of a pulse via the meteor and the station named.

An example of digital recording is given in Fig. 9.12.

9.10   Foundations of the Method

Figure 9.13 gives a simplified flow chart of the data-processing method for wind-profile determination.

The amplitude and phase of the Fresnel pattern generated by a meteor of constant magnitude, without diffusion, can be expressed in terms of the Fresnel integrals of classical analysis (e.g., McKinley, 1961); but numerical integrations are necessary to predict the effects of diffusion or varying magnitude (Loewenthal, 1956; Southworth, 1962). When, as with our stations, the *range* (distance to the meteor trail) is much larger than the distance between the transmitting and the receiving stations, it is convenient and entirely adequate to use the concept of an "effective station." This is the position that would receive the same signal by direct back-scatter that the actual receiving station receives by forward-scatter at a small angle; it is located on the bisector of the angle between lines joining the specular reflection point with the transmitting and the receiving stations, and its range R from the meteor is the harmonic mean of the transmitting and receiving ranges. The position and speed of the meteor are deduced primarily from the times of the maxima and minima (collectively, *extrema*) of the Fresnel pattern; diffusion, radar magnitude, and other physical quantities are deduced primarily from the amplitudes of the extrema. The

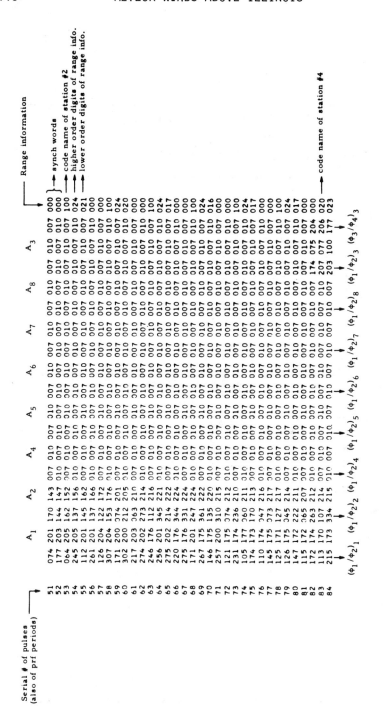

Fig. 9.12   Example of digital recording (from a meteor simulation run).

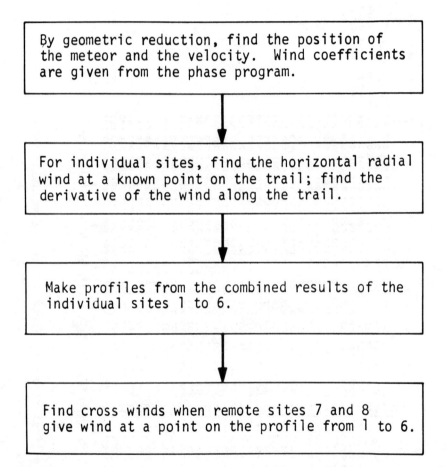

By geometric reduction, find the position of the meteor and the velocity. Wind coefficients are given from the phase program.

For individual sites, find the horizontal radial wind at a known point on the trail; find the derivative of the wind along the trail.

Make profiles from the combined results of the individual sites 1 to 6.

Find cross winds when remote sites 7 and 8 give wind at a point on the profile from 1 to 6.

Fig. 9.13 Simplified flow chart of wind-profile computer programs.

times of the extrema are those when the meteoroid is at known
multiples of the length F of the *principal Fresnel zone* from
the specular reflection point. From the numerical integrations
cited above, we find that the *known multiples* depend on apparent diffusion, but not much on other likely events. Diffusion
shifts maxima earlier and minima later than in a pattern without diffusion. However, the time of the first maximum also
depends on the slope of the ionization curve (Southworth,
1962, Table 3). From the integrations, an empirical expression
for the distance x_1, from the specular reflection point to
the point corresponding to the first maximum of the Fresnel
pattern, is

$$F(0.861 - 1.535C + 2.75C^2 - 3.0C^3 + CS/\sqrt{2})/\sqrt{2} \qquad (9.1)$$

where S is the slope of the ionization curve in magnitudes per
length F, and C is Loewenthal's *decay constant*:

$$C = \frac{8\pi D}{\lambda 2_V} \sqrt{\frac{\lambda R}{2}} = T_F/2\sqrt{2\pi} \; T_D \qquad (9.2)$$

Here, D is the diffusion constant, V is the meteor's velocity,
T_F is the time for the meteor to cross the principal Fresnel
zone, and T_D is the time for the voltage amplitude of the
signal to decay by a factor e under diffusion.

Analytical developments by Southworth (1962) show that
the amplitudes of the Fresnel patterns may be quite accurately
analyzed by smoothing out the oscillations and then regarding
the smoothed curve as denoting the decaying amplitude from the
principal Fresnel zone and the oscillations from the smoothed
curve as denoting the amplitudes of the later Fresnel zones
as each is formed. Such an analysis yields C, S, and the
radar magnitude at the specular reflection points and at the
extrema. It also yields the phase corrections necessary for
wind measurements in the Fresnel pattern.

The *radiant* (named for the point in the sky from which
visual shower meteors appear to radiate) is the direction
opposite the direction of motion of a meteor and is the
common way to denote the direction of motion. Evidence for
the radiant is found in the time intervals between crossings
of different specular reflection points, seen as time intervals
between beginnings of Fresnel patterns. The distance traveled
by the meteoroid in a given time interval is known from the
velocity (found from the spacing of Fresnel oscillations).
This distance is the projection of the distance between the
corresponding effective stations onto the meteor trajectory.
Given at least two such projections from noncolinear inter-
station distances, the radiant is determined.

The position in space of the specular reflection point from station 3 (the transmitting station) is determined from the radiant, the range (directly measured), and the difference in phase between the signals received at the two halves of the antenna at station 3. The difference in phase determines the angle between the direction from one-half of the antenna to the other and the direction to the meteor. There is an ambiguity between two or three possible values of the angle, which can nearly always be resolved by requiring the meteor to be in the antenna beam and at a plausible height; any doubtful cases are discarded.

The reduction procedure begins by treating the range observations, i.e., the observed time intervals between the transmission of a pulse and its return to station 3 via the meteor, or via the meteor and another station. Discrepant values are discarded, and one average loop range formed for each station with range measures.

The observed (amplitude) Fresnel patterns are measured by a specially devised semiempirical pattern-recognition program. This finds the time and amplitude of each extremum, thus giving the same data as we previously read from analog film records. The program uses the spacing of extrema found at one station to predict the spacing at others. If it fails in an attempt to measure the pattern at any station, it will try again when it gets a better prediction from some other station. The oscillations from a smoothed curve are found for each pattern.

The wind phase Φ is next measured at each station for which an amplitude pattern was measured. The recorded phase data are an analog representation of s = amplitude χ sin (phase) at odd pulses and c = amplitude χ cos (phase) at even pulses. (These are the outputs of the phase detectors.) Initially, the missing alternate values are interpolated. For a first approximation to the phase variation, the phases are found at the Fresnel maxima, multiples of 2π being added where necessary for continuity. These are then fitted with an expression of the form

$$\Phi = E_0 + E_1 p + E_2 p^2 \ , \tag{9.3}$$

where p is the pulse number (measuring time). Next, the phase at each observed pulse after the first maximum is corrected for the oscillating part of the Fresnel pattern by subtracting a rotating vector

$$s_{corr} = s - C_F \sin(\phi_F + E_0 + E_1 p + E_2 p^2)$$

$$c_{corr} = c - C_F \cos(\phi_F + E_0 + E_1 p + E_2 p^2)$$

$$\tan\phi_{corr} = s_{corr}/c_{corr} \tag{9.4}$$

Here, C_F is the amplitude of the oscillating part, inter-polated between extrema; and ϕ_F is the phase of the oscillating part, defined to be 0 at the first extremum, π at the second, 2π at the third, etc., also interpolated. The corrected phases are fitted with (9.3), and the process then iterated once more. If E_2 is inadequately determined or significantly positive, it is rejected, and

$$\Phi = E_0 + E_1 p \tag{9.5}$$

is fitted instead.

The amplitude Fresnel pattern is next analyzed at each station by using a provisional value of the range. This yields velocities and times at specular reflection points, for the geometric reduction proper, as well as various physical data such as diffusion.

The geometric reduction proper (determining the position and velocity) follows next; it proceeds by iteration. First, the effective station positions are estimated. Next, the velocities and times at specular reflection points are corrected. Then, the velocities V and times t are fitted to an expression of the form

$$V = B + C\bar{K} \exp(\bar{K}t) \tag{9.6}$$

Whipple and Jacchia (1957) have shown that this form is suitable for the analysis of photographic (Super -Schmidt) meteors. Since it does not yet appear practical to attempt to evaluate $\bar{K}$ from individual radar meteors, we use a value characteristic of faint photographic meteors - namely, 5/4 of the value expected for unfragmenting meteoroids in an exponential atmosphere. Accordingly,

$$\bar{K} = 1.25 \ \bar{V} \cos Z_R/5.3 \tag{9.7}$$

where $\bar{V}$ is the mean observed velocity, Z_R is the zenith distance of the radiant, and the atmospheric scale height is taken to be 5.3 km. We do not impute great accuracy to (9.6), but consider that it is a more reasonable bridging formula than most others, especially polynomials. The distances between specular reflection points are now found by integrating

(9.6), and the radiant is fitted to these and the effective
stations. Finally, the specular reflection point from station
3 is found by fitting the position of the whole trail (not
varying the radiant) to the difference of phase at station 3
and to all observed ranges.

Successive iterations begin by computing the effective
stations, using the newly computed position of the trail, and
continue by correcting velocities for the difference between
the original provisional ranges and the latest values. The
iteration is carried to convergence (or failure).

Throughout the reduction, it is general practice to *fit*
by least squares and to carry an estimate of the uncertainty
of nearly every quantity deduced. This estimate is kept as
realistic as possible and usually depends on both the internal scatter in a least-squares fit and the previously estimated uncertainties of the data fitted. In case of doubt,
we try not to underestimate the errors.

Figure 9.14 gives a flow chart of the processing applied
to phase information, and Fig. 9.15 is a more detailed chart
of the wind-profile computer programs.

The machine time required by these computations is
approximately equal to the recording time, when a CDC 6400 is
utilized. This is why we claim that the SAO system is poten-
tially capable of real-time wind-pattern measurements.

9.11 Wind Profiles from Individual Meteors

Winds are computed in a rectangular coordinate system
whose origin is at sea level, 100 km horizontally from station
3 in the direction of the antenna beam - namely, 113°east of
true north. The Z coordinate is measured vertically from
that point; the X coordinate, horizontally in the prolongation
of the antenna beam; and the Y coordinate, horizontally at
right angles to X and Z, in a generally north-northeast direction. The great majority of our wind data are collected within, say, 30 km horizontally from the origin of these coordinates. *Horizontal* winds are defined as winds parallel to the
X-Y plane; thus, they may be up to about 1/3 degree from the
true local horizontal. *Radial* winds are defined as winds in
a vertical plane containing station 3, positive to the east
and southeast. "Cross" winds are defined as winds in a vertical plane normal to a vertical plane containing station 3,
positive to the north and northeast. However, when there are
insufficient data to determine both radial and cross-wind

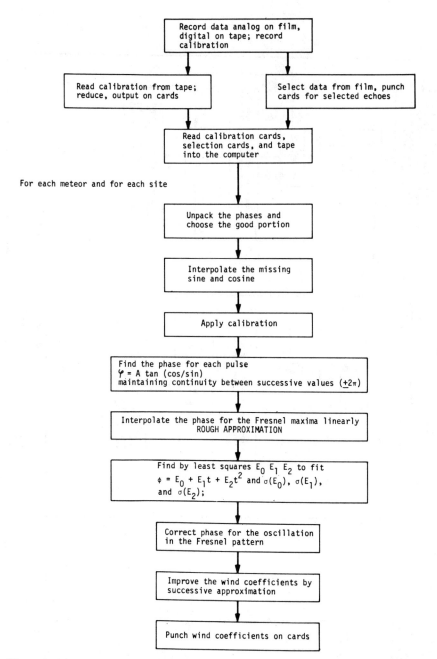

Fig. 9.14 Flow chart of the processing of the phase information. Strategy: Resolve the Fresnel oscillations into a smoothly decreasing component of constant phase. When the rotating component is subtracted, the phase of the remaining echo can give the best values for the phase shift due to wind.

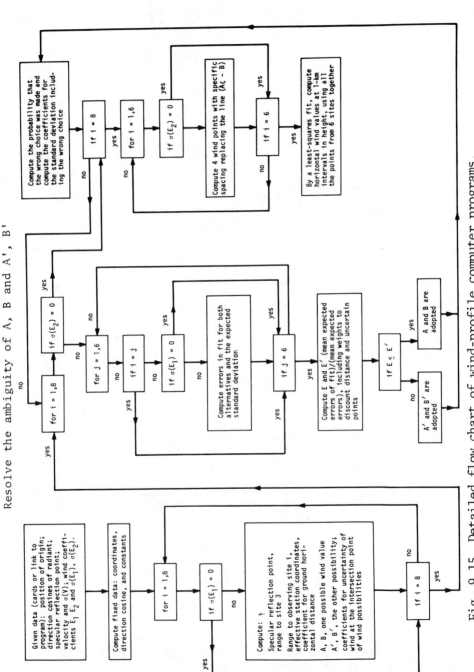

Fig. 9.15 Detailed flow chart of wind-profile computer programs.

components at a given point, any wind component measured at
stations 1 to 6 will be regarded as a radial component. This
coordinate system has been chosen to be suitable for the data
in an intermediate stage of processing, not for examination of
the results. When there are sufficient data, the final results
can be expressed in any convenient scheme.

In computing a wind profile from an individual meteor, we
invariably assume that there is no vertical component to the
wind and transform all observed wind components to horizontal
wind components on that assumption. However, we retain information for studying vertical components by combining more than
one meteor; this is described further on.

The quantities used in computing a wind profile are the
meteor trajectory (one point and the radiant), the corrected
positions of the specular reflection points and the corrected
velocities at those points (both corrected for wind), the
phase coefficients E_0, E_1, E_2 (equation 6.3) at each station,
and expected errors of all these. The phase coefficients are
reduced to wind components. When there is an ambiguity at a
station (when E_2 was accepted), it is resolved by comparing
the errors of fit of both alternatives to the unambiguous wind
values. Here we compute both the sum of the squared errors and
the sum of squared errors expected from the prior estimates
of the wind uncertainties. Furthermore, we give high weight
to errors within the portion of the trajectory occupied by
the principal Fresnel zone during the wind-phase measurement
and rapidly decreasing weight outside that portion; the two
alternatives correspond to different portions of the trajectory.
Comparing the observed errors with the expected ones, we estimate the separate probabilities of the alternatives. The
more likely alternative is then chosen, but its estimated
squared error is increased by the squared difference between
the alternatives multiplied by the relative probability of the
rejected alternative.

A profile of horizontal radial wind is now constructed by
using all observations from stations 1 to 6. Each linear
"profile" observed at a particular station is replaced by four
points lying on the linear profile and spread over the part of
the trajectory occupied by the specular reflection point during wind-phase measurement. Then, at each integral kilometer
in height within the part of the trajectory where winds were
measured, a cubic polynomial in distance along the trajectory
is fitted to whatever wind points are near that height. The
weights assigned these points depend on their estimated uncertainties and decrease rapidly for points away from the height
in question. The tabular wind at each height is then taken

from the cubic centered at that height and these values are used to develop the wind profile in the vertical. In the next step for each tabular value of horizontal radial wind, we also compute the interpolated effective station position. This defines the actual direction of the measured radial wind component, as averaged in the process of combining data from different stations to make one profile. This "interpolated station" is computed by finding the effective position of each station and then fitting these positions with a linear expression in distance along the trajectory at the same time and with the same weights, as a cubic is fitted to the winds (described above).

One or more values of horizontal cross wind are computed when there is a reliable wind measure (E_2 accepted) from station 7 or 8 at a height within the profile of horizontal radial wind and within the portion of the trajectory occupied by the principal Fresnel zone from that station during wind-phase measurement. At each such height, we have two horizontal wind measures, one from the effective position of station 7 or 8 and one from the interpolated effective station position for that point on the profile. These are combined vectorially to yield the radial and cross components, and the newly found radial component replaces the less exact value in the profile.

To determine vertical wind components, we take advantage of the baseline represented by the separation of our stations 1 and 2 from stations 5 and 6, as well as the two baselines between 7 or 8 and our remaining stations. When the baseline 1-2-5-6 is used, we still need to have two meteors nearly at the same point at nearly the same time, but we need to have measures from station 7 or 8 on only one of the meteors. For this purpose, we compute interpolated station numbers, corresponding to interpolated effective station positions, for effective positions between stations 1 and 2 or 5 and 6. We use these numbers to mark suitable wind values and to convey the effective station positions to the computation that combines data from different meteors.

9.12 Wind Field Computed from Many Meteors

The bulk of our wind observations occur in a volume defined horizontally by the antenna-gain pattern and vertically by the occurrence of meteor trails. The horizontal extent is, roughly, 50 km in the X direction and 30 km in the Y direction; the vertical extent is roughly 20 km. At peak rates, several hundred meteors per hour are observed in this volume. The computer can readily average and smooth all these data, but we must first decide upon the resolution in space and time.

A vertical resolution of 1 km has been chosen because:
1) it seems appropriate to the scale of wind irregularities
observed by rocket wind measures, 2) it is approximately the
accuracy of the radar height measures (so that a finer resolution would fail), and 3) it is approximately the height resolution in the individual radar wind profiles. Here we estimate the resolution in individual profiles from the vertical
separation of adjacent specular reflection points, which is
always less than 3 km and has a mode near 1 km, and from the
loss of any resolution within the principal Fresnel zone,
which has a vertical extent approximately from 0.1 to 0.5 km.
The vertical resolution is also now fixed by our data formats.

Unlike the vertical resolution, the horizontal and temporal resolutions are not yet fixed, and the computer program will use whatever resolution we ask. With one or two
thousand radial wind values per hour (at 1-km vertical resolution), we can resolve a few hundred cells in space-time.
The natural scale for horizontal variation seems to be a hundred or more kilometers, and, for temporal variation, to be an
hour or perhaps less. If we try for fine spatial resolution,
we can select, say, a 10-km horizontal interval and a 20-min
time interval; if we look for fine time resolution, we might
choose a 30-km horizontal interval and a 1-min time interval.
At less than peak measurement rates, we should not choose resolutions this fine, and it will always be necessary to use
lower resolution for cross winds.

The computation of the wind fields begins with the radial-
and cross-wind values computed from many meteors in a defined
interval of time. Each 1-km interval in height is treated independently. The tabular intervals in horizontal position and
time are specified in advance for each computation. At each
point in the X-Y time grid, a value of horizontal radial wind
is computed if there are six or more measures within the eight
neighboring cells in X-Y-time space. The value is found by
fitting a linear expression in X, Y, and time to the measures
and taking its value at the tabular point. Similarly, a value
of cross wind is computed if there are six or more cross wind
measures. For convenience in later use, missing tabular values
are then interpolated if there were at least six adjacent tabular values determined. The interpolation uses the same fitting
procedure, starting with the tabular values rather than with
measures; interpolated values are marked as such.

9.13   Three-Dimensional Wind Measures

Whenever in the course of tabulating the wind field we
find that any one space-time cell·contains either cross-wind

measures from both stations 7 and 8 or a cross-wind measure
effectively from stations 1-2 or 5-6 and a radial-wind measure
from 5-6 or 1-2, respectively, a vertical wind component will
be computed. First, the error in relative height between the
two meteors will be eliminated as accurately as possible by
matching their horizontal radial profiles. Then, all the relevant observed components will be transformed back to the original radial observations from different effective stations and
combined vectorially (by least squares where possible) to find
three components of the wind. More advantageous solutions
will be possible when more than two meteors are available.

9.14 Expected Wind-Measurement Accuracies

It is certain that the accuracy of measurement will vary
widely from one measure to another, and the computer programs
have been carefully planned to give as realistic an estimate
of the errors as possible in each case. Here we will attempt
only to predict some typical accuracies.

The basic phase accuracy per pulse will be of the order
of 0.05 to 0.1 rad, limited mostly by noise and digitizer resolution for weak pulses and by saturation and Fresnel corrections for strong pulses. In a typical observation lasting
0.1 sec (74 pulses) with only partly random errors, the phase-
rate accuracy is of the order of 0.3 rad/sec, corresponding
at our wavelength to a radial wind of 1.0 mps and a horizontal
radial wind of 1.2 m/sec.

Interpolation of the wind between specular reflection
points (construction of a wind profile) adds considerable error,
depending on the second and higher derivatives of the wind with
height, but perhaps typically 3 mps.

Cross-wind measures always use interpolated wind values
and must be computed by combining two vectors at an angle of
about 0.15 rad. Consequently, their typical error will be 20
mps.

Combination of the data from different meteors to evaluate
the wind field will tend to increase the apparent errors of
the horizontal radial wind because the errors in height (1 km)
combined with typical wind shears (0.01/sec) will appear as
errors in the wind (10 mps). However, the errors in tabular
mean values will be smaller, perhaps 5 mps. Combination of
cross winds, when there are sufficient data, will tend to decrease the errors to 10 to 15 mps.

The typical error in a vertical wind component when only two meteors are used will be similar to the error in a single cross wind. Interpolating errors in two matched profiles should partly cancel to give a relative error of 2 m/sec, but this must be divided by an angle of about 0.1 rad. Combining four meteors, the typical error would perhaps be 10 m/sec.

9.15 Examples of Results

Figure 9.15 shows a few reconstructed profiles of horizontal radial wind. The downward progression of the profiles with time is suggestive of the presence of a gravity wave.

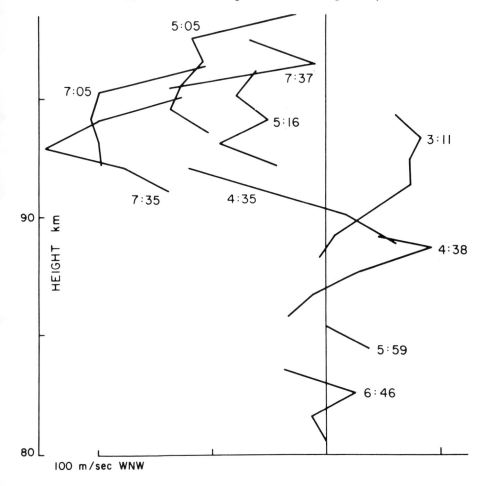

Fig. 9.15 Wind profiles over Decatur, Illinois, on the morning of August 25, 1969 (labeled in Central Standard Time).

9.16   References

Deegan, N. F., R. J. Fitzpatrick, G. Forti, M. D. Grossi, M. R. Schaffner, and R. B. Southworth, 1970. "Study of meteor wind measurement techniques." Final Report on Contract AF 19(628)-3248, AFCRL-70-0168(I) and (II), February.

Grossi, M. D., 1963. "High altitude wind measurements by collecting and processing meteor radar echoes." Conference on Direct Aeronomic Measurements in the Lower Ionosphere, Conference Record, University of Illinois, Urbana, Ill., pp. 82-88, December.

Loewenthal, M., 1956. "On meteor echoes for underdense trails at very high frequencies." Lincoln Lab. Tech. Rep. No. 132, Massachusetts Institute of Technology, Cambridge, Mass., December.

Manning, L. A., O. G. Villard, and A. M. Peterson, 1950. "Meteoric echo study of upper atmosphere winds." Proc. IRE, Vol. 38, pp. 877-883.

McKinley, D. W. R., 1961. Meteor Science and Engineering. McGraw-Hill Book Co., New York, 309 pp.

Southworth, R. B., 1962. "Theoretical Fresnel patterns of radio meteors." Harvard-Smithsonian Radio Meteor Project Res. Rep. No. 14, 60 pp.

Southworth, R. B., 1966. "Height and wind accuracy: the Havana System." AFCRL-68-0228, Spec. Rep. No. 75, pp. 161-169.

Whipple, F. L., and L. G. Jacchia, 1957. "Reduction methods for photographic meteor trails." Smithsonian Contr. Astrophys., Vol. 1, pp. 183-206.

CHAPTER 10

THE STANFORD METEOR RADAR SYSTEM

A. M. Peterson and R. Nowak

Stanford University, California

10.1   Introduction

The measurement of winds by radar observation of the drift
of meteor trails was first undertaken at Stanford *two decades*
ago. It was soon realized that processing of the data produced by the radar was a complex and time consuming process by
the methods then available. Recently, in cooperation with the
Meteorology Laboratory of the Air Force Cambridge Research
Laboratories (AFCRL), techniques have been investigated at
Stanford for *automated* collection and digital signal processing
of radar data from meteor trails.

As a result of the techniques study, a new pulse-doppler
radar and digital data acquisition system has been designed and
constructed. The primary aim of the system is to obtain wind
data as a function height on a continuous, relatively *unattended*
basis.

10.2   The Stanford Meteor Radar System Design Criteria

Considering the limitations on meteor-radar measurements
a worthwhile undertaking was believed to be the study of oscillations and circulations in the upper atmosphere, possibly
using several widely separated stations .for synoptic measurements. An earlier pilot system (Nowak, 1967) was constructed
which produced good results at low cost. Atmospheric density
measurements also were found promising, and the previously
described *height calibration* method was developed for application in wind measurements and for continuous monitoring of
density fluctuations. Under the constrains of low cost, easy
mobility, and simple rugged construction for unattended operation (as expected from one system in a network of stations),
not all the possibilities of meteor-radar measurements can be
exhausted. For example, multiple-station arrangements that
allow the direct measurement of turbulence were excluded.

In the early work with meteor trails, echoes were commonly recorded on film and then reduced by hand. The amount of data generated even by a fairly low-power system made operation for more than short stretches prohibitively expensive and, as a result, one of the first specifications for the present system was the reduction of data by computer. Because experiments with analog-recording techniques revealed that the subsequent conversion to digital form for acceptance by the computer was overly expensive, digital recording on *magnetic tape* was chosen. This type of recording is not novel, even for meteor work; for example, the AFCRL system (Barnes and Pazniokas, 1968) employs a small on-line digital computer. The Stanford system, however, has found an exceptionally simple solution where data are recorded essentially in raw form in the format illustrated in Fig. 10.1. Only a relatively small amount of digital circuitry is required, and reduction of the data is left completely to a large computer that can do this task very efficiently.

For the Stanford radar a number of factors led to the use of a *pulse doppler system*, rather than a C.W. doppler system. Foremost among these is the difficulty of operating a C.W. system in the presence of large numbers of aircraft. Experience at Stanford has shown that echoes from aircraft can occur as much as 50 percent of the time. As discussed below, a high duty cycle, high average power pulse transmission was chosen in order to maximize echo occurrence rates and minimize transmitter cost. An unusual phase-coded pulse ranging system was designed which permits good range accuracy while using long pulse transmission intervals.

From Manning and Eshleman (1959), the number of trails with a line density of q or larger is inversely proportional to q. Echo power is proportional to q^2 and transmitted power. Therefore, the number of echoes detected above a certain given threshold will be proportional to the square root of the transmitted power. As expected, data rates depend on transmitted power; however, this dependence is not very strong. For a given receiving system, if the data rate is to be doubled, the transmitted power must be quadrupled. Peak power does not completely describe the sensitivity of the system. In fact, average rather than peak power is the important parameter.

The quantity of interest for the sensitivity of the system is the signal-to-noise ratio at the output of the receiver. At the typical frequencies of meteor-radar, sky noise and man-made interference are the principal sources of noise. Interference can be diminished by a careful choice of operating frequency and a suitably protected location. Sky noise, however, is appreciable so that low-noise design of the receiver

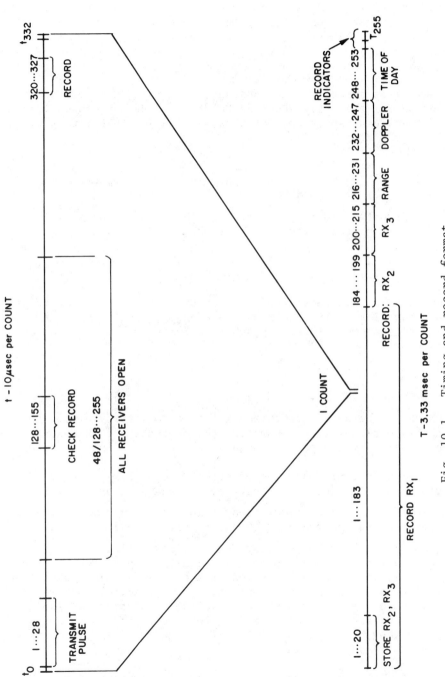

Fig. 10.1 Timing and record format.

is not critical; in a normal receiver, the signal-to-noise ratio
at the output is essentially the same as at the antenna. Over
a reasonable bandwidth, the noise power is

$$P_N \propto kTB \qquad (10.1)$$

where k is the Boltzmann constant, T is the "sky temperature"
(at 30 MHz typically about 30,000°K), and B is the bandwidth
of the signal spectrum; for a pulse with length t_p, this width
is

$$B = \frac{2}{t_p} \qquad (10.2)$$

    From the radar equation (McKinley, 1961), the signal-to-
noise ratio of the received pulse can be derived as

$$\frac{S}{N} \propto \frac{P}{B} \qquad (10.3)$$

where P is the transmitted peak power, and the constant of pro-
portionality depends on radar wavelength, range, target size,
and noise temperature. By substituting B from above, one obtains

$$\frac{S}{N} \propto Pt_p \qquad (10.4)$$

which is the energy contained in one transmitted pulse. For
the same signal-to-noise ratio, therefore, the peak power can
be lowered if the pulse length is increased at the same time.

    With a limit on P (here established at 5 kW), it is ad-
vantageous to choose as long a pulse length t_p as possible;
however, another consideration establishes a maximum value for
t_p. Because the transmitter and receiver are located at the
same site, the receiver must be inhibited during transmit time
(and during some additional time to allow for switching tran-
sients). With echoes desired at a certain minimum range, the
receiver must open soon enough after the start of the transmitted pulse so that a pulse traveling to a target at this
range and back can be received. For the Stanford system, a
pulse length of 280 µs, with a 120 µs guard time, yields a
minimum range of 60 km.

    For the measurement of the echo amplitude, each pulse can
be treated individually, and the pulse repetition rate need be

just high enough to provide adequate sampling of the echo
envelope. For phase measurements (e.g., determination of the
doppler), however, as high a repetition rate as possible is
desirable, not only to remain above the Nyquist rate for the
sampled waveform but also to increase the amplitude of the
waveform reconstituted from the samples. The maximum pulse
repetition rate is limited by the maximum range at which the
signals are to be received. One design goal was the reception
of meteor echoes at elevation angles above 15°. This value is
a compromise between the desire to measure the horizontal winds
at as small an angle as possible and the requirement of a high
pulse repetition rate. For an angle of 15°, a meteor trail at
100 km height will present a range of 375 km, corresponding to
a round-trip time of the signal of 2.5 ms. With additional
time needed between a receive period and the next transmitted
pulse, for signal processing, a pulse repetition rate of 300
pps was chosen.

In Nowak (1967), a detailed discussion on the choice of the
optimal signal frequency of a meteor radar is presented. Echo-
power and decay-time constants are the principal quantities
depending on frequency, but many others are also functions of
frequency (among them the diameter of the first Fresnel zone
which constitutes the radar target and over which the measured
quantities appear averaged). From the considerations given
(Nowak, 1967), the optimal value is between 30 and 40 MHz. The
exact choice in this band will depend on the availability of a
frequency assignment and, in normal cases, it can be assumed
that such an assignment is secured in advance of operations and
that the assigned frequency is available over a prolonged period
of time. In the interest of simplification, the Stanford system
was designed to operate at a fixed frequency. Because network
operation with the stations assigned different frequencies is
envisioned, an attempt was made to minimize the number of
modifications required to change the operating frequency. By
appropriate partitioning of the system, only one printed circuit
board in the main frame needs to be changed completely; on
other boards, some slight adjustment of component values might
be necessary to permit proper tuning.

Among other parameters, the angle of arrival of the signal
must be determined to obtain azimuth and target height. In
principle, four methods are available for this measurement. In
the classical radar technique, an antenna with a split beam and
a pronounced null in the pattern along its axis is moved until
the signal disappears in this null; the antenna then is aligned
with its axis pointing toward the target, and its attitude provides the measured angle. In the case of meteor echoes, which
are short and rapidly decaying, the antenna cannot move fast

enough mechanically, and electronic beam steering is required. In the preliminary design of the Stanford system, this technique was believed to be too complicated for reasonable accuracy (an angle measurement to within ±1° in elevation is required to obtain meaningful height measurements). Recently, however, such a method has been demonstrated successfully (Rudmann, 1969), and its adaptation to a system similar to the one described in this report is being studied.

In the second technique, triangulation is employed; i.e., the ranges to the trial from several ground stations are measured. This method was shown (Nowak, 1966) to be poorly suited for meteor trails because, on a line target, the re-flection points for different stations are at different loca-tions and large errors result in the analysis of the data.

The third technique utilizes the fact that the wavefront of the incoming signal is essentially plane, at right angles to the direction of propagation. If two antennas are lined up at an angle with the wavefront, the signals will be shifted in phase with respect to each other, and the amount of phase shift will depend on azimuth and elevation angles of the echo. Phase comparison among three antennas will yield both angles indi-vidually.

In the fourth technique, the angle of arrival of the sig-nal can be measured by comparing signal strengths at two an-tennas with different radiation patterns. The elevation angle can be obtained from two antennas with equal patterns in azimuth but different patterns in elevation (realized by placing iden-tical antennas at different heights above ground). Azimuth could be measured in a similar manner, with two antennas whose patterns differ in azimuth only. This configuration is more difficult to implement and does not need to be used because, with two antenna pairs, elevation and azimuth can be determined individually even if each pair has radiation patterns that de-pend on both elevation and azimuth. The only requirement would be that the two pairs of antennas are chosen in such a way that their families of curves of constant signal ratio in the azimuth-elevation plane intersect at nearly right angles so that errors are minimized.

These last two methods were extensively investigated and compared by Nowak (1966), and the conclusion reached was that both methods should yield about the same angle accuracy, ±1°. Because the pattern-comparison technique is easier to implement in a pulsed system, it was chosen for the Stanford station. Three receivers are provided; one is used for measurement of echo amplitude (decay), range, and doppler shift, and each of

the other two is switched between two antennas, during successive receive periods, to provide pattern comparison. Two
pairs of antennas thus are sampled to obtain both azimuth and
elevation angles.

Experience with the system has indicated that, in the pattern-comparison method, the accuracy of ±1° should be obtainable
for elevation-angle measurements above 30° (to exclude site
effects) using dipoles at different heights above a ground well
defined by a wire mesh acting as a perfect reflector. To date,
no firm quantitative results on the actual accuracy have been
achieved because calibration runs could not provide a good
standard to compare with the antenna measurements. Calibration
procedures, involving a satellite beacon and aircraft tracked
by high precision radar, are now being used, and it is anticipated that they will provide reliable results concerning the
accuracy of the angle measurement.

It has been noted that, to achieve high sensitivity of the
system at moderate peak power, the transmitted pulse must be
of fairly long duration. This precludes ranging with normal
radar techniques; because range resolution is directly proportional to bandwidth (i.e., inversely proportional to pulse
length), the long pulse employed would yield range with completely inadequate accuracy. Another method must be employed,
therefore, and the one chosen was phase modulation of the
transmitted pulse with a pseudo-random code (Nowak, 1966;
Golomb, 1964). In this technique, as it was first applied in
radar, the phase of the transmitted signal is switched between
-90° and 90° in a certain pattern. This pattern (code) then
can be recovered at the receiver by a simple phase detector.
The output of this detector is compared with the original code.
If the code is chosen properly, its autocorrelation function
has one sharp peak at 0 and a small value at any other position;
i.e., when the received signal is correlated (every bit-time)
with the original code, a large output results only when they
are perfectly "lined up." This is the point at which the complete pulse has returned to the receiver; from the time at
which coincidence occurs, the range of the echo is obtained.
Resolution is one bit-time of the code. In the Stanford system,
with 28 code bits of 10 μs duration, the resulting range resolution is 1.5 km (or, if the center of the bit is taken as
reference, ±0.75 km).

In Golomb (1964), the spectrum for a code-modulated pulse
is derived and shown to be similar to an unmodulated pulse
train with a pulse length of one code bit. Considering that,
in the correlation, the bit amplitudes are added coherently
and noise is added incoherently, the signal-to-noise ratio for

ranging would be that of a 10 µs pulse of 28 times the peak power of the actual 280 µs pulse. For digital correlation, employed here, the signal-to-noise ratio is degraded by 3 dB (Spira, 1966).

The bandwidth corresponding to a 10 µs pulse (200 kHz) is unacceptable for amplitude measurements because an improvement of the signal-to-noise ratio similar to the ranging case is not present. The coding, therefore, is done by switching the phase of the signal only between -45° and 45°, which corresponds to the superposition of a phasor switched between -90° and 90° and a constant phasor of equal magnitude at 0°. Because the latter corresponds to an unmodulated pulse of the full duration of 280 µs, its bandwidth is only approximately 7 kHz, and the signal-to-noise ratio is comparable to that in the ranging.

The signal, thus, is split into two parts; half the energy is contrained in a narrow band in which amplitude is measured, and half is in a wide band for ranging. The precision of the switching in phase is not critical, however. If the phase change is not exactly 90°, the superposition of a constant phasor and one switching between 0° and 180° is still a valid description of the situation, but the two phasors are no longer of equal magnitude.

10.3 Application of the Stanford Radar System

Because of the probable use of this radar in a network for *synoptic* measurements, the design was specifically tailored for easy reproducibility at low cost. Construction is on printed circuit boards for which negatives are available, and the system is partitioned in such a way as to isolate different measurements as much as possible so that modifications normally would not involve more than one circuit board at a time. A block diagram of the system is shown in Fig. 10.2.

The following functions are available.
  (1) Generation of a transmitted pulsed signal with a peak power of 5 kW minimum. The transmitted pulse can be coded for ranging.
  (2) Three receivers: one measures peak amplitude during each receive interval, doppler shift, and range; the other two measure peak amplitude only and can be switched between two antennas on successive pulses.
  (3) Conversion of data into digital form, intermediate storage, and formating for recording on digital magnetic tape in fixed length records.
  (4) A digital clock displaying time in hours, minutes, and seconds, recorded with each echo.

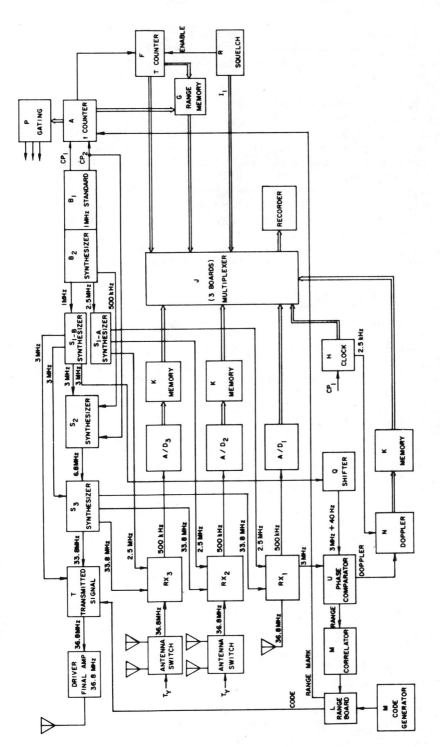

Fig. 10.2.   Block diagram of the Stanford system.

The complete station, including a cathode-ray oscilloscope for testing and a communications receiver, fits into two standard 19-in. racks of approximately 52-in. height. A photograph of the radar (except for antennas) is shown in Fig. 10.3.

This radar system can be used for a variety of measurements. If the transmitted pulse is not code modulated, no range can be obtained, but the full peak power is available for amplitude and doppler measurements. In this mode of operation, height is not determined directly but is derived from echo decay using a decay-height relationship established independently. Thus, measurements can be made at low-elevation angles to take advantage of high data rates. If an antenna with a fairly narrow beam (of, say, 50° width) and small sidelobes is utilized, no individual measurements of either elevation or azimuth angles are necessary; on the average, the wind component in the direction of the beam axis is obtained (Nowak, 1967).

In this configuration, only one receiver is used and, with a TR switch, it can share the antennas with the transmitter. Two antennas can be employed pointing in orthogonal directions (NW and NE are the directions for highest data rates), and the transmitter and receiver are switched between them at regular (hourly, for example) intervals. The prototype Stanford system was used for quite some time in this mode. Antennas were 3-element Yagis mounted 0.6λ above ground, with a resulting beam elevation of approximately 30°. Beamwidths were about 60° in the horizontal and 50° in the vertical.

Height resolution is not reliable in this mode; a standard deviation of ±2 km in the measured heights was quoted above. Conversely, data rates are high because of operation at low-elevation angles, resulting in smaller errors in the average values obtained.

In the full-measurement program, amplitude, doppler, range, azimuth, and elevation angles can be measured simultaneously. At present, the station is used in such a program at Eglin AFB, Florida. The transmitter antenna is a 3-element Yagi pointed north, as is the antenna for receiver 1, from which amplitude, decay, doppler, and range are obtained. Receiver 2 switches between two collinear dipoles at λ/2 and 3λ/16 above ground, respectively, and elevation angle is determined from this measurement. Receiver 3 switches between two Yagis pointed NE and NW, respectively, to provide information on the azimuth angle. In these antennas, the signal ratio is not dependent solely on azimuth; however, because elevation angle was determined independently, azimuth can be obtained. It should be noted that, in this arrangement, wind and height are measured

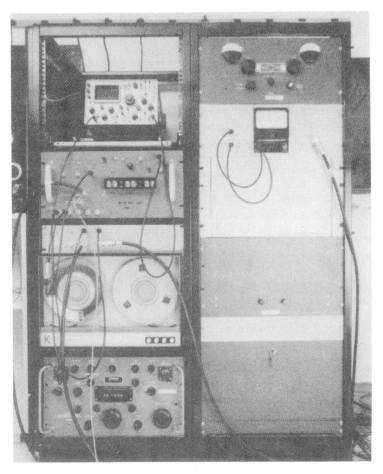

Fig. 10.3 Complete MTR station. Left rack, from bottom:
          communications receiver, tape recorder, antenna
          coupler, main frame, oscilloscope. Right rack:
          driver and final amplifier with their power supplies.

at the same time. Wind measurements should be made at low-
elevation angles, and sufficient height accuracy is obtainable only at high ones; therefore, operation around 45° is indicated as a compromise, and all Yagis are inclined to achieve
maximum antenna gains in that direction.

10.4 Conclusions

The Stanford system is adaptable and has proven to be
simple to maintain, reliable, and useful in a variety of
measurement programs. Negatives for all printed circuit boards

are available, and construction of another set should not present any difficulties.

10.5 References

Barnes, A. A., Jr. and J. J. Pazniokas (editors), 1968. "Proceedings of the Workshop on Methods of Obtaining Winds and Densities from Radar Meteor Trail Returns." Special Report No. 75, AFCRL-68-0228, Air Force Cambridge Research Laboratories.

Golomb, S. W., (editor), 1964. Digital Communications, Prentice Hall, Inc., Englewood Cliffs, New Jersey.

Manning, L. A., and V. R. Eshleman, 1959. "Meteors in the Ionosphere." Proceedings of the Institute of Radio Engineers, Vol. 47, 186-199.

Nowak, R., 1966. "Height Determination of Radar Echoes from Meteor Trails." Report SU-SEL-67-001, Stanford Electronics Laboratories, Stanford University, Stanford, California.

Nowak, R., 1967. "An Integrated Meteor Radar System for Wind and Density Measurements in the Upper Atmosphere." Final Report, SU-SEL-67-046, Stanford Electronics Laboratories, Stanford University, Stanford, California.

Nowak, R., M. North, and M. S. Frandel, 1970. "The Stanford Meteor-Trails Radar Mark II." Report SU-SEL-70-021, Stanford Electronics Laboratories, Stanford University, Stanford, California.

Rudmann, R. J., 1969. "Azimuth Angle Subsystem for Meteor Trails Radar." Term paper EE 891, Department of Electrical Engineering, University of New Hampshire, Durham, New Hampshire.

Spira, P. M., 1966. "Digital Detection of Binary Coded Radar Signals." Report SU-SEL-66-001, Stanford Electronics Laboratories, Stanford University, Stanford, California.

CHAPTER 11

INTERACTIONS BETWEEN THE NEUTRAL ATMOSPHERE
AND THE LOWER IONOSPHERE

C. F. Sechrist, Jr.

University of Illinois,
Urbana, Illinois

11.1 Introduction

Several types of ionospheric measurements have revealed *anomalous* D-region behavior during winter, and corresponding effects have been observed in airglow emissions originating at D-region heights (Bowhill, 1969a). The correlations found between changes in the D region and stratospheric temperatures are of particular interest (Bossolaso and Elena, 1963; Shapley and Beynon, 1965). Correlations between features of anomalous D-region behavior and planetary-scale features on high-level weather maps must be investigated. Limited knowledge of the morphology of D-region anomalies seems to be the major obstacle to such a study.

Studies of the electron density distribution during a few days of anomalously large medium-frequency (~ 2 MHz) radio-wave absorption (winter anomalous days) have shown increased electron densities in the D-region, particularly above 80 km, and partial-reflection observations during winter have also shown a marked day-to-day variability at lower heights. Although studies in Europe have suggested a lower-latitude limit for the occurrence of the winter anomaly, little information is available about the general morphology of the anomalous D-region condition (Lauter, et al., 1969).

Atomic oxygen, nitric oxide and water vapor are minor constituents in the D region that have long photochemical lifetimes, and it is therefore expected that *dynamical processes* will have pronounced effects on their distributions and consequently on the electron density distribution. Models incorporating eddy diffusion have demonstrated that this is the case. Relatively modest vertical velocities, however, can produce comparable transport. Unfortunately, the nature of the dynamical processes cannot be specified at this time.

261

Present knowledge of the *transport processes* themselves is rather limited. Numerous studies have shown that part of the spectrum of internal gravity waves excited in the troposphere can propagate to D-region heights without damping. The problem at present is to identify the sources, in the troposphere or elsewhere, which force motion on the space and time scales of internal gravity waves. The sources of planetary-wave energy are, on the other hand, comparatively well understood; any 10-mb weather map adequately defines the lower boundary condition for a full-wave solution of the planetary-wave equations. The problem is that these waves are strongly refracted in the wind field between 30 and 70 km and are subject to damping as well, so that accurate solutions require a precise knowledge of atmospheric structure in this region. It is now clear from theoretical studies (Dickinson, 1968; 1969) that planetary waves can propagate to the D region in winter with less attenuation than in summer; the same may be true of gravity waves, depending on their height of origin.

11.2 Evidence for Meteorological Influences on the Lower Ionosphere

There is now convincing evidence which suggests that electron densities in the lowest ionosphere (the D region between 50 and 90 km) are influenced by or associated with the temperature and wind structures of the neutral atmosphere. The associations between the neutral stratosphere and the lower ionosphere have been described by several investigators, and the existence of these associations is usually considered to constitute the evidence for *stratosphere-ionosphere coupling* (SIC). In the early studies of interactions between the neutral and ionized atmospheres, the experimental state of the art was such that very little was known about meteorological parameters above 30 km (balloon radiosonde altitude limit), and medium- and high-frequency (2-4 MHz) radio wave absorption measurements were used to monitor indirectly changes in the D-region electron density. Thus, it should not be surprising that the evidence for stratosphere-ionosphere coupling was obtained earlier than the evidence for mesosphere-ionosphere coupling.

The bulk of the evidence for SIC phenomena has been provided by radio wave absorption and balloon temperature results. Radio measurements using frequencies in the VLF, LF, MF, and HF ranges have been used successfully to detect the effects of the neutral lower atmosphere on the lower ionosphere. The

majority of the SIC studies appear to have used 10-millibar
temperature data, because for many years this was the upper
limit of meteorological data.

Subsequent to radio absorption studies there have been
studies involving radio *drift* measurements, partial reflection studies of D-region electron densities, rocket-grenade
measurement of temperatures, rocket measurement of electron
densities, and airglow measurements.

Mechtly and Smith (1968) reported *seasonal variations*
of the lower ionosphere at 38°N geographical latitude
(Wallops Island, Virginia) during years of minimum sunspot
number. Essentially, rocket electron density profiles were
measured at 60° solar zenith angle in each season. As shown
in Fig. 11.1, summer solstice and equinox electron densities
were found to be larger than those of winter solstice by a
factor of about four between 65 and 80 km. Because the
measurements were conducted during the years of the quiet
sun (IQSY) it is reasonable to assume that the variations
in electron density profiles were caused by chemical and/or
transport processes in the mesosphere and were *not ascribable*
to fluctuations in solar ionizing radiations.

Mechtly and Shirke (1968) presented rocket electron
density profiles for the *normal* and *anomalous* winter days of
December 15, 1965 and January 10, 1966, respectively, when
the solar zenith angle was 60°. Enhancements of electron concentration were found on the anomalous day relative to the
normal day, from 60 to 95 km altitude, ranging up to a factor
of 47 at 85 km, as shown in Fig. 11.2. This enhancement in
electron density was shown to explain the increased 3 MHz
radio wave absorption observed by ground-based instrumentation. Figure 11.3 illustrates the behavior of the average
(1000-1200 EST) 3.03 MHz radio absorption measured at Wallops
Island, Virginia, during the period January 26 to February
3, 1967. Note the gradual buildup in absorption over a
period of several days, the well-defined peak on January 31st,
and the gradual decay of absorption. One of the outstanding
results of these measurements was the fact that the bulk of
the *anomalous* radio absorption occurred above 80 km, as shown
in Figs. 11.4 and 11.5. Thus, the implication of this
result is that only the positive-ion chemistry need be considered in an aeronomical study of the winter anomaly. Consequently only those minor constituents which influence the
electron production rate and the positive-ion chemistry need
be scrutinized for their sensitivity to temperature changes
and changes in transport processes.

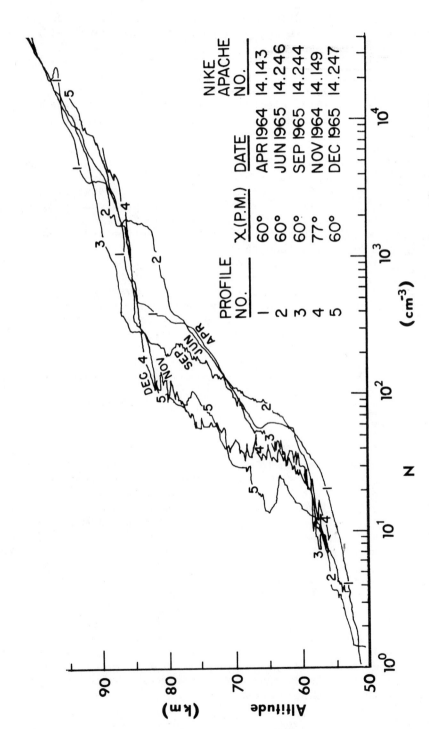

Fig. 11.1 Seasonal variation of rocket electron densities for 60° solar zenith angle (Mechtly and Smith, 1968).

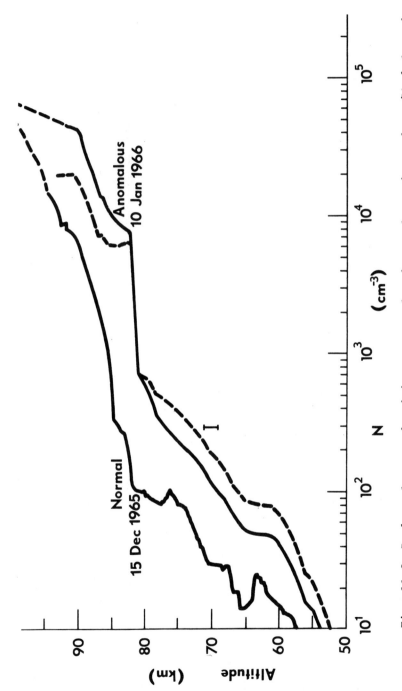

Fig. 11.2 Rocket electron densities on normal and anomalous winter days (Mechtly and Shirke, 1968).

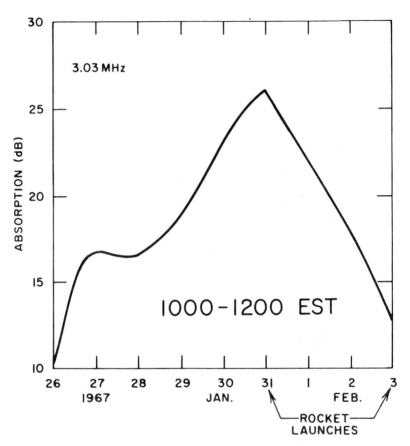

Fig. 11.3 Average 3.03 MHz radio wave absorption at Wallops
Island, Virginia (Mechtly and Shirke, 1968).

Because of possible associations between D-region warmings and enhanced electron densities on winter days of high
radio-wave absorption, nearly simultaneous rocket measurements of D-region temperatures and electron densities were
conducted on an anomalous winter day at Wallops Island,
Virginia (Sechrist, et al., 1969). The electron density profile for the anomalous day (January 31, 1967) was generally
similar to the one measured on the anomalous winter day of
January 10, 1966, as shown in Fig. 11.6. The rocket-grenade
temperature profile for the anomalous day revealed a strong
temperature inversion at 70 km and a warm layer extending
from the inversion to 84 km. The temperature structure was
significantly different at the same altitudes on the normal
winter day of February 3, 1967. It was concluded by Sechrist,
et al. (1969) that a meteorological hypothesis of the origin

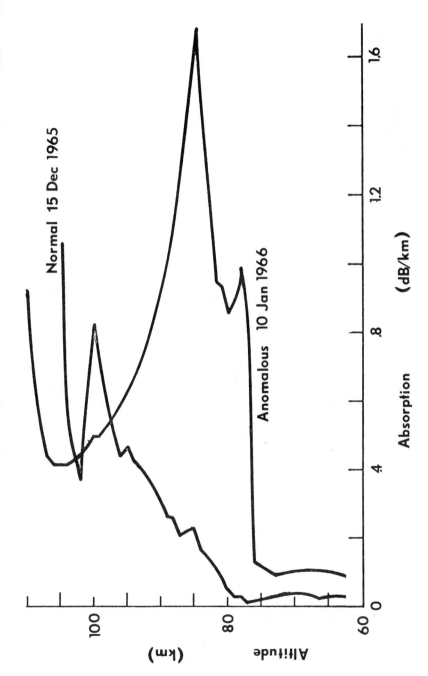

Fig. 11.4 3.03 MHz absorption rate calculated from rocket electron density profiles (Mechtly and Shirke, 1968).

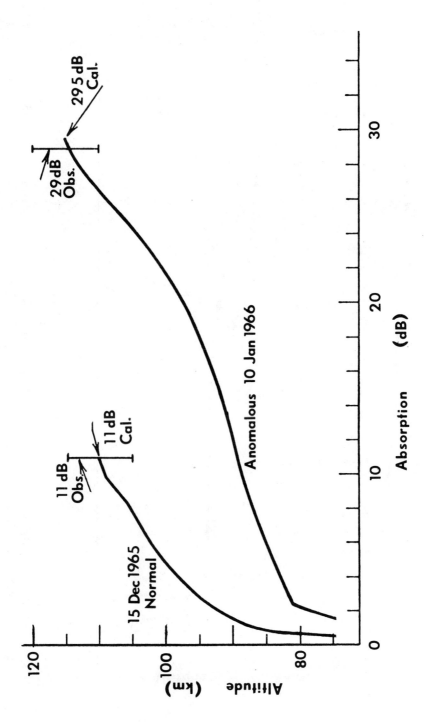

Fig. 11.5 Total 3.03 MHz absorption profiles (Mechtly and Shirke, 1968).

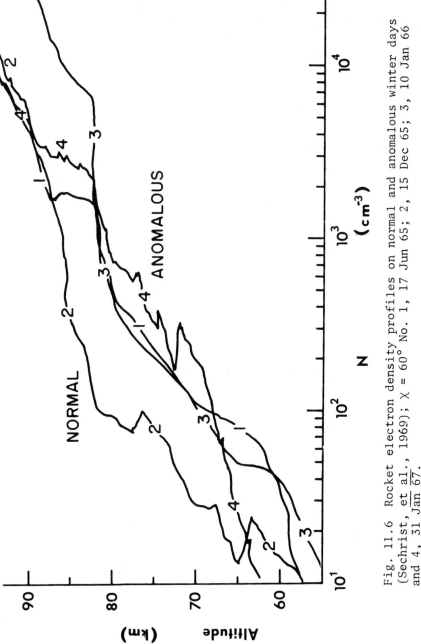

Fig. 11.6 Rocket electron density profiles on normal and anomalous winter days (Sechrist, et al., 1969); χ = 60° No. 1, 17 Jun 65; 2, 15 Dec 65; 3, 10 Jan 66 and 4, 31 Jan 67.

of the D-region winter anomaly was strengthened by the results of the above-mentioned coordinated rocket experiments.

Temperature structures on the normal and anomalous winter days of February 3, 1967 and January 31, 1967, respectively, were significantly different; this can be seen in Fig. 11.7. These temperature profiles are characteristic of the middle latitudes in winter (Nordberg, et al., 1965; Theon, et al., 1967). Mesospheric soundings conducted over a period of several hours at Wallops Island have indicated (Theon, et al., 1967) that the irregular features in the temperature and wind structure of the mesosphere are transient in nature, as shown in Fig. 11.8. An interesting result of the rocket-grenade measurements was the observation that variability in summer seemed to decrease with increasing latitude, and the wave-like temperature structure was suppressed entirely at Barrow, Alaska.

The temporal and spatial behavior of the D-region winter anomaly in MF and HF radio wave absorption is perhaps the outstanding example of evidence for interaction between the neutral atmosphere and the ionosphere. Thomas (1962) examined the world morphology of abnormally high absorption in winter months at middle latitudes for the northern hemisphere, using measurements of radio wave absorption and supplementary f_{min} data obtained during the IGY. Thomas found that the condition of very high absorption on any day extended over an area of at least 10^6 square kilometers. Also, this study revealed that days of enhanced anomalous absorption in one longitude sector corresponded to days of reduced anomalous absorption in another sector. Dieminger (1969) has pointed out that the correlation is almost perfect over distances of a few 100 kms, but that it drops to 0.6 or 0.7 over distances of 1000 km. Although Thomas (1962) found an anticorrelation between the occurrence of anomalous winter days over North America, Europe and Asia, respectively, Dieminger, et al. (1968) found a striking correlation between electron density profiles measured in Canada and radio wave absorption as measured at Lindau, Germany on individual days.

Appleton and Piggott (1954) considered the anomalous behavior of ionospheric absorption in winter. They noted that enhanced absorption was found in winter on a relatively high proportion of days. Furthermore, if Appleton and Piggott selected, for each of the winter months which exhibit the anomaly, 20% of the least absorbed days instead of all days, the anomalous behavior disappeared; i.e., the absorption on these days compared well with the summer values when

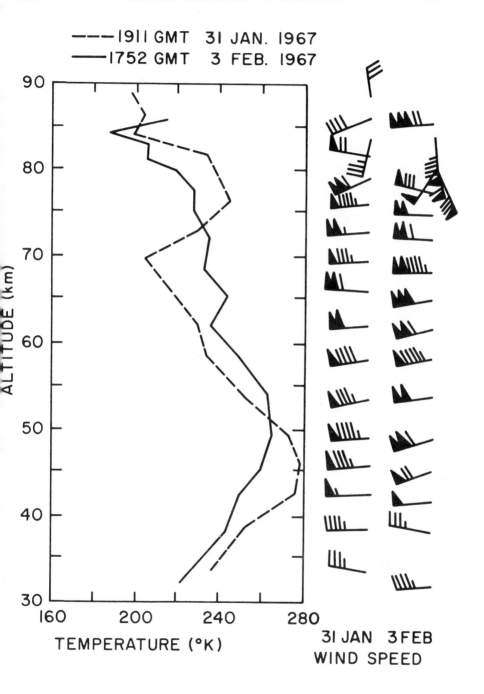

Fig. 11.7 Rocket grenade temperature and wind profiles on normal and anomalous winter days (Sechrist, et al., 1969).

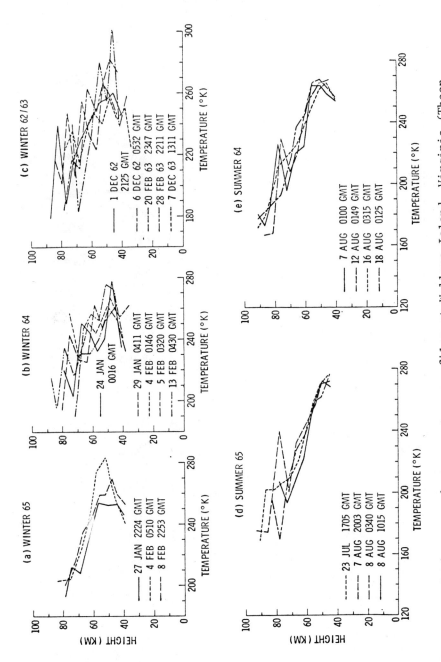

Fig. 11.8 Rocket grenade temperature profiles at Wallops Island, Virginia (Theon, et al., 1967).

normalized to the same solar zenith angle. This suggested
that for some winter days the same absorption mechanism was
operative as in summer, and that the anomalously high absorption on the other winter days was due to some mechanism which
increases the basic absorption already present. Anomalous
winter absorption is present on single days or on groups of
days between November and February in the Northern hemisphere;
and one of the main properties of the winter anomaly is the
strong interdiurnal variability.

The winter anomaly is most distinct in middle latitudes.
It increases in strength towards the winter polar region but
becomes obscured by auroral and polar absorption events in
the higher latitudes. Towards the equator, the winter anomaly
decreases gradually between 40° and 35°; and it becomes un-
detectable at latitudes below approximately 30°. Lauter and
Schaning (1970) discovered a definite low-latitude boundary
of the winter anomaly phenomenon at 37°N by means of measurements on board a ship on a route from Europe to Central
America. Dieminger (1952) noted that, on groups of days in
winter, anomalously high absorption was accompanied by weak
partial reflections from below 100 km; this implied that
electron densities were enhanced down to 90 km and below.
Beynon and Davies (1955) showed that the anomalous absorption
involved a non-deviative component at mesospheric altitudes,
and some evidence suggested corresponding changes in electron
densities at E-region altitudes. Lauter and Nitzsche (1967)
demonstrated that the major portion of the effect was located
between 80 and 100 km, and Gregory (1956) showed that increased absorption on particular days was accompanied by increases in the amplitudes of partial reflections from 66 to
88 km. Also, Gregory and Manson (1969) confirmed the conclusion that the height region of the enhanced electron
density in winter must be situated above 80 km by means of
partial reflection experiments.

According to rocket measurements of electron density
(Mechtly and Shirke, 1968; Sechrist, et al., 1969) the winter
anomaly in absorption is caused predominantly by enhanced
electron densities in the upper D region above 80 km approximately. Measured changes in electron collision frequency
are inadequate to account for the anomalously high absorption
on certain winter days. Also, Thomas (1968) deduced electron
density enhancements in the D and E regions during days of
anomalous radio wave absorption in winter by means of absorption and virtual height measurements of MF and HF radio
waves.

The winter anomaly in absorption tends to be associated

with other ionospheric parameters. For example, the *ionization drift* around 95 km reverses its direction on certain days in winter, and this may be associated with wind reversals at stratospheric altitudes. Also, the maximum electron density of the E layer is *higher* in winter than in summer when reduced (normalized) to constant solar zenith angle.

Correlations between ionospheric and stratospheric phenomena are impressive but not straightforward. Although there is a coincidence between very strong peaks of stratospheric temperature and excessive radio absorption, the correlation is not unambiguous. In fact, the breakdown of the winter polar vortex may sometimes be a better indicator of anomalous lower ionosphere electron densities.

It should be mentioned that not all winter days of anomalous ionospheric absorption can be associated with changes in some neutral atmospheric parameter. For example, in recent years several workers (Maehlum, 1967; Belrose and Thomas, 1968; Bourne and Hewitt, 1968; Manson, 1968; and Manson and Merry, 1970) have shown that statistically the D-region electron density enhancements are correlated with magnetic storms. However, the enhancement is not on storm day but is delayed by 3 to 4 days after the storm at middle latitudes; and this phenomenon has been labelled the magnetic storm *after effect*. Hence, it appears that recent work has made less clear the previously reported correlation between ionospheric absorption and certain neutral atmospheric parameters. In particular, means must now be devised to enable one to separate anomalous winter effects caused by meteorological factors from those due to energetic particle precipitation effects.

In summary, the D-region winter anomaly may be described by the following list of characteristics:

1. There is enhanced MF and HF radio wave absorption on single days or groups of days at middle latitudes in winter.

2. The observed anomalous ionospheric radio absorption has a diurnal trend and midday maximum similar to that for a "normal" winter day, except that absorption values are significantly larger.

3. Anomalous ionospheric absorption may be localized geographically; i.e., it may be observed over an area of at least 10^6 km^2.

4. Partial reflections from the ionospheric D region may be present.

5. Negative associations or anticorrelations may exist between different longitude sectors.

6. Anomalous absorption may be associated with major or minor warming events at stratospheric levels.

7. Enhanced D-region electron densities may be associated with pressure ridges in the stratosphere, the implication being that vertical motions redistribute ionizable constituents.

8. Anomalous absorption appears to be caused by enhanced electron densities in the 80 to 100 km altitude region, and this implies that E-region critical frequencies may be higher than normal.

9. Anomalous absorption may be associated with magnetic storm *after effects*, implying that enhanced electron densities are caused by energetic particle (electron) precipitation effects.

10. Anomalous absorption may occur simultaneously or nearly simultaneously with abrupt changes in ionization *drifts*, suggesting that polar vortex reversals and instabilities are possible at mesopause altitudes and above.

11.3 Aeronomy of the Upper D Region

This section describes the present knowledge of electron production and loss processes in the ionospheric D region between altitudes of roughly 75 km and 90 km. This height region is of interest because it is now known to be responsible for the anomalous winter absorption of ionospherically reflected radio waves. Apparently, the height distribution of one or more minor neutral constituents is responsible for the steep ledge observed in rocket electron density profiles around 82 km. It is suggested that winter electron density enhancements between roughly 80 and 90 km may be caused by changes in the relative abundance of simple molecular positive ions (NO^+ and O_2^+) and hydrated positive ions of the form $H+\cdot n(H_2O)$ where $n = 1,2,3,\ldots$. This notion is strengthened by the fact that positive water-cluster ions are known to possess electron-ion recombination coefficients which are larger than those for simple molecular ions (Reid, 1970a).

Before one can ascertain the cause or causes of the mid-latitude D-region winter anomaly, it is important to determine the factors which influence the electron density in the middle and upper D region; this is because anomalous radio wave absorption is now known to be caused by anomalous increases in the electron density between 80 and 90 km, approximately.

In a complete theoretical study of the undisturbed, mid-latitude, daytime D region, one must know what chemical and transport processes determine the electron density. Certain ground-based experiments (partial reflection and wave interaction) have been used to deduce electron density distributions in the D region, and rocket measurements of electron density are now rather commonplace. Rocket measurements of positive and negative ion density have been conducted by several experimenters, but there are problems associated with the interpretation of these data. Positive-ion composition in the D region has been measured by rocket ion mass spectrometers since 1963 (Narcisi and Bailey, 1965), and the first negative ion composition measurements were made in 1969 (Narcisi, et al., 1970).

Presently, there are theoretical models of positive and negative ion density and composition; however, these are usually *revised* within months after being presented because of important new discoveries in laboratory studies and/or ion composition measurements by means of rockets under various ionospheric conditions. Over the last two or three years, numerous theoretical and laboratory studies of ion-molecule reaction schemes for D-region positive and negative ions have been carried out. Equally important have been many theoretical and laboratory investigations of the chemistry of the minor neutral constitutents several of which influence strongly the electron density.

Unfortunately, experimental techniques for measuring the neutral species composition are few. Attempts have been made thus far to measure the concentrations of NO, O, $O_2(^1\Delta)$, O_3, and H_2O with varying degrees of success. Ironically, we seem to know more about the electron density and ion composition than the concentrations and composition of minor neutral constituents.

Our knowledge of the minor neutral constituent transport processes in the upper mesosphere (D region) is even more meager than that of the minor neutral species concentration and composition. Apparently, both chemical and transport processes play important roles in determining the concentrations and distributions of the minor neutral species.

Much is unknown about the relevant chemical reaction rate constants, recombination coefficients, and their tempera- ture dependences; however, impressive progress in laboratory measurements of aeronomic interest has been made during the last three or four years, and future laboratory studies are expected to yield a wealth of new data on both neutral and ion-molecular reactions.

Clearly, there are certain neutral species having chemi- cal lifetimes (time constants) large compared to the time constants for transport; and these neutral species are re- distributed by means of the transport process. Hence, we need to know considerably more about the *meteorology* of the D region before it will be possible to synthesize realistic theoretical models for the minor neutral constituents and, consequently, for the positive and negative ion species.

The electron density of the D region is best discussed starting with the continuity equation for electrons:

$$\frac{d\{e\}}{dt} = P - L$$

where $\{e\}$ denotes electron density in electrons cm^{-3}, P is the electron production rate in electrons cm^{-3} sec^{-1}, and L is the electron loss rate. These production and loss terms change with altitude in the region between 50 km and 100 km (Reid, 1970a; Sechrist, 1970).

Let us first consider the production of electrons. Be- low about 65 km, the principal ionization process is the ionization of all the atmospheric constituents by galactic cosmic rays. It is thought that the ionization of nitric oxide by Lyman α radiation is the major source of ionization up to about 90 km. Recently, it was suggested by Hunten and McElroy (1968) that the ionization of $O_2(^1\Delta_g)$ by solar UV radiation between 1027 and 1118Å was another important source of electrons in the D region. Since then, however, Huffman, et al. (1971) have shown that CO_2 absorption in the O_2 win- dows is serious and that ionization of $O_2(^1\Delta_g)$ is probably not a competitive process with Lyman α photo-ionization of NO. Above 85 km the major ionization process is by 2 to 8 Å X rays and by Lyman β radiation near the bottom of the E region. This assumes the absence of particle precipitation effects, and these effects will be discussed separately.

D-region electrons are subject to two primary loss mechanisms: dissociative recombination with molecular or water-clustered positive ions, and attachment to neutral species to form negative ions. It is thought that the most important loss process for electrons below about 75 km is electron attachment primarily to O_2 molecules. Between about 75 and 85 km, it has been suggested that the principal electron loss process is dissociative recombination with the water-clustered ions 19^+ $(H^+ \cdot H_2O)$, $37^+ (H^+ \cdot 2\{H_2O\})$, $55^+ (H^+ \cdot 3\{H_2O\})$, etc. (Sechrist, 1970; Reid, 1970a). Above this level, say from 85-100 km, electrons are lost mainly through dissociative recombination with NO^+ and O_2^+.

Dynamic processes are thought to be of great importance in the D region. This does not result from the direct transport of electrons (due to their short lifetime), but rather to the dynamical influences upon the minor neutral constituents that enter into the electron production or loss processes either directly or indirectly. That is to say, if a constituent that does not enter directly into electron production or loss acts to control the concentration of a constituent that does enter into these reactions, it is very important in determining electron density. An example to be discussed later is the possible importance of H_2O concentration on electron density in the middle D region.

In the lower D region (below 65 km), where both the electron production and loss processes work through major constituents, the important influence of dynamics would be to increase or decrease the atmospheric density since the electron concentration can simply be shown to be inversely proportional to the atmospheric density at a given altitude.

Between 65 and 90 km the electron density should be strongly dependent on the concentration of nitric oxide. Since NO is produced in the E region (Norton and Barth, 1970; Strobel et al., 1970), any dynamical process that enhances or suppresses vertical transport in the D region would be very important.

The region between 75 and 85 km should be under the strongest dynamic control in the D region if the present ideas concerning electron loss by recombination with hydrated ions are correct. This would be for two reasons: First, Ferguson and Fehsenfeld (1969) have suggested that the relative concentrations of the water-clustered species (19^+, 37^+, 55^+, etc.) are a very sensitive function of temperature, and Reid (1970a) has suggested that the recombination coefficients of these ions increase markedly with the increased complexity of these ions. This being the case, increases in temperature between 75 and 85 km would be expected to result in an in-

creased electron density. In addition, Sechrist (1970) has
suggested that the concentration of these water-clustered
ions should be a function of the water vapor concentration.
Thus, it might be expected that changes in the water vapor
transport from below would have a marked effect upon the
D-region electron density.

Finally, we come to the role of energetic particles in
the D region. Maehlum (1967) has suggested that *precipitating energetic electrons* may be responsible for the winter
anomaly in ionospheric absorption. In this work, he used
satellite observations of particles and found that the energy
content of such corpuscular radiation at midlatitudes was
sufficient to produce an absorption anomaly similar to that
which is observed. There have been rocket measurements of
precipitating particles by a Russian group (see Tulinov, 1967,
and Tulinov, et al., 1969) suggesting that, although energetic
particle precipitation is a principal ionizing agent during
the night, it is of lesser importance than solar radiation
during the day at middle and low latitudes. Manson and Merry
(1970) and Potemra and Zmuda (1970) have performed detailed
calculations supporting this view.

The neutral species which enter into the positive ion
chemistry of the ionospheric D region are reviewed next;
these species include NO, $O_2(^1\Delta)$, H_2O, and O. Only the
positive ion chemistry of the D region is considered in this
paper because it is known that the altitude region above about
80 km is responsible for anomalous winter absorption of MF
and HF radio waves; and negative ions are not important above
about 75 km in the daytime D region, as demonstrated by Reid
(1970a) and Sechrist (1970).

Minor neutral species such as O, NO, $O_2(^1\Delta)$, and H_2O
play a major role in influencing the electron density through
photoionization and/or ion-molecule processes. For example,
positive ions and free electrons result from the photoionization of NO and $O_2(^1\Delta)$, and all four constituents are involved
in several important ion-molecule chemical reactions.

A minor neutral constituent that plays a dominant role
in D-region ion chemistry is nitric oxide (NO). Nicolet
(1945) originally suggested that NO must be an important
source of D-region ionization because its relatively low ionization potential allows it to be photoionized by solar Lyman
α radiation at 1216Å. The reaction $NO^+ + h\nu(1216A) \rightarrow NO^+ + e$
is generally recognized as providing most of the daytime ion-
ization that exists during quiet solar conditions in the
70 to 90 km height range. Rocket measurements of the NO

concentration have been carried out by Barth (1966a, 1966b), Pearce (1969), and Meira (1970).

The mechanism for production of NO from nitrogen atoms was for many years believed to be (Nicolet, 1965)

$$N + O_2 \rightarrow NO + O + 9.14 \text{ kcal} \tag{11.1}$$

with a 7.1 kcal activation energy, which was then balanced by

$$N + NO \rightarrow N_2 + O + 134.6 \text{ kcal} \tag{11.2}$$

Reactions (11.1) and (11.2) together have the property that the equilibrium concentration of NO is independent of the concentration of atomic nitrogen (N), if chemical equilibrium prevails; it gives the concentrations shown in the first column of Table 11.1. Rocket measurements of the NO concentration by Barth (1966) and by Pearce (1969) from the fluorescence of the NO γ-bands gave much larger values, as shown in Table 11.1 also. However, it should be noted that observations by Meira (1970) have pointed to lower values for the NO concentration, and Pearce (1970) has revised his earlier values downward by almost a factor of 10, because of a correction for Rayleigh scattering.

Table 11.1   Comparison of observed and calculated D-region nitric oxide concentrations in cm^{-3}

Altitude (km)	Nicolet (1965)	Barth (1966)	Pearce (1969)	Meira (1970)	Strobel, et al. (1970)
90	2×10^5	5×10^7	10^8	2×10^7	5×10^7
80	1.4×10^6	5×10^7	2×10^8	1.5×10^7	10^8
70	5×10^6	----	10^9	6×10^7	----

Norton and Barth (1970) suggested that another source of NO might involve excited nitrogen atoms. The reaction

$$N(^2D) + O_2 \rightarrow NO + O \; 144 \text{ kcal} \tag{11.3}$$

is rapid and, unlike (11.1), has no activation energy. The consequences of this reaction have been explored by Norton and Barth (1970) and by Strobel, et al. (1970), who find it is generally adequate to match both D- and E-region measurements of NO concentration.

In Fig. 11.9, the NO concentration profile deduced by Pearce (1969) is compared with that derived from earlier rocket measurements by Barth (1966). Pearce's measurements

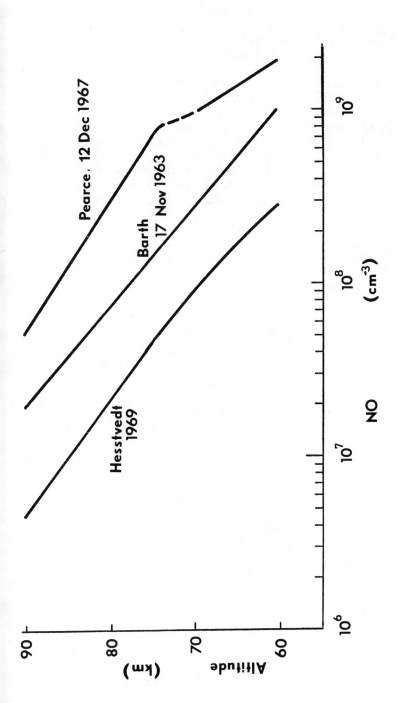

Fig. 11.9 Comparison of theoretical and experimental nitric oxide distributions (Sechrist, 1970).

show that NO concentrations are roughly a factor of 4 greater
than Barth's values in the vicinity of 80 km, assuming a mixed
distribution for NO passes through Barth's experimental value
of 3.9 x 10^7 cm^{-3} at an altitude of 85 km.

Another NO distribution shown in Fig. 11.9 is based on
theoretical work by Hesstvedt and Jansson (1969) who considered the effect of vertical eddy transport on the distribution of NO in the D region.   They showed that vertical eddy
transport modifies the photochemical models for NO in a drastic way and brings the theoretical results into much closer
agreement with Barth's experimental result.   Figure 11.10
shows NO profiles deduced by Meira (1970) from two rocket
experiments at Wallops Island in early 1969.

$O_2(^1\Delta)$ is another dominant minor neutral constituent
in the mesosphere, and it is important in D-region aeronomy
because it can be photoionized by a solar UV band (1027-1118Å)

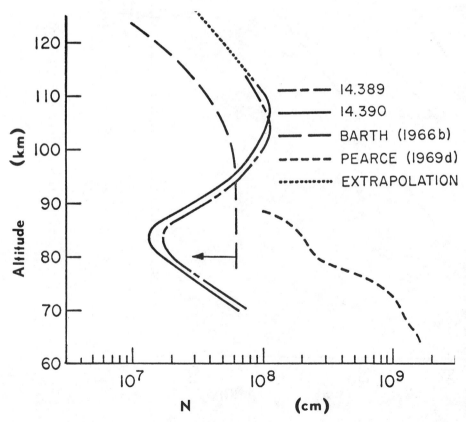

Fig. 11.10 Comparison of experimental nitric oxide distributions (Meira, 1971).

to produce O_2^+ ions and free electrons. The O_2^+ ions thus produced play important roles in the positive-ion chemistry of
the D region.

The presence of substantial concentrations of $O_2(a^1\Delta_g)$
molecules has been inferred from twilight and dayglow studies
of the infrared atmospheric bands at 1.27μ and 1.58μ, and reliable altitude profiles have been obtained by measurements
with rocket-borne infrared photometers (Evans and Llewellyn,
1970). The concentration attains a daytime maximum value
approaching 3×10^{10} cm^{-3} at an altitude near 50 km.

The first measurement of the dayglow height distribution
of the 1.27μ O_2 infrared atmospheric band was made by Evans,
et al. (1968) with a rocket photometer flown at White Sands
(33°N). A combination of the observed height distribution
and the diurnal variation of total intensity measured with
balloon-borne photometers (Evans, et al., 1969) led to the
conclusion that the dominant mechanism for the production of
$O_2(^1\Delta)$ is the photolysis of ozone as originally proposed by
Vallance-Jones and Gattinger (1963).

While the ozone photolysis mechanism is satisfactory in
accounting for the height distribution of $O_2(^1\Delta)$ up to 70 km,
the predicted concentrations are lower than those observed
above this height. The height distribution obtained by Evans,
et al. (1968) seemed to show a definite upper layer of $O_2(^1\Delta)$
at 85 km.

$O_2(^1\Delta)$ is one of the key measurable quantities related
to the photochemistry of the mesosphere because of its importance in producing O_2^+ ions and free electrons, and in the
associative detachment of O_2^- negative ions ($O_2^- + O_2(^1\Delta) \rightarrow$
$2O_2 + e$).

Summer daytime height profiles of $O_2(^1\Delta)$ concentration
were measured at Fort Churchill by Wood, et al. (1970). The
height distributions showed main peaks at 55 km (solar elevation of 18.2° in the afternoon) and at 59 km (solar elevation of 9.2° in the morning). The $O_2(^1\Delta)$ concentrations above
80 km showed significant enhancements over those predicted
from ozone photolysis on the basis of the ozone concentrations
predicted by most previous theoretical models. Height distributions of $(O_2(^1\Delta))$ derived by Wood, et al. (1970) and
by Evans, et al. (1968) are shown in Fig. 11.11.

Recently, Megill, et al. (1970) have shown that the
ozone concentration can be deduced from the $O_2(^1\Delta)$ measurements
in the daytime mesosphere. Based on the two reactions for

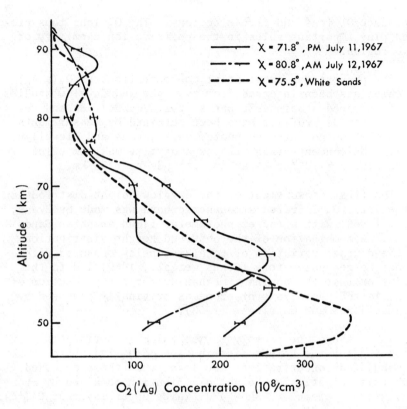

Fig. 11.11 Experimental distributions of $O_2(^1\Delta)$ (Wood, et al., 1970).

the production and loss of $O_2(^1\Delta)$:

$$O_3 + h\nu \rightarrow O_2(^1\Delta) + O(^1D) \qquad (11.4)$$

$$O_2(^1\Delta) \rightarrow O_2 + h\nu \qquad (11.5)$$

it is easily shown that

$$\{O_2(^1\Delta)\} \stackrel{\sim}{\sim} 34 \{O_3\} \qquad (11.6)$$

Reaction (11.4) is the direct result of ozone photolysis in the Hartley band, while reaction (11.5) refers to the quenching of $O_2(^1\Delta)$ known to take place down to 50-60 km. Megill, et al. (1970) have recently reported on very significant $O_2(^1\Delta)$ measurements in the aurora.

Ground-state atomic oxygen, $O(^3P)$, is important in both mesospheric positive-ion and negative-ion chemistry. In the

upper D region, O plays an important role because of the reaction

$$O_4^+ + O \rightarrow O_3 + O_2^+ \tag{11.7}$$

Essentially, reaction (11.7) competes with the reaction

$$O_4^+ + H_2O \rightarrow O_2^+ \cdot H_2O + O_2 \tag{11.8}$$

which is an important link in the chain of hydrated positive
ion-molecule reactions postulated by Ferguson and Fehsenfeld
(1969). In the lower daytime D region (below $\sim$ 75 km), atomic oxygen is important in the associative detachment of O_2^-
negative ions through the reaction

$$O_2^- + O \rightarrow O_3 + e \tag{11.9}$$

Because of their interdependence and the relationship
between certain of their ionospheric influences, O, O_3 and
H_2O should be considered together. Hunt (1966) was the first
to include time-dependent solutions for ozone and hydrogen
compounds; however, he neglected transport effects altogether.

Hesstvedt (1968) studied theoretically an oxygen-hydrogen model of the mesosphere and lower thermosphere, including
the effects of vertical eddy transport. Refer to Figs.
11.12 and 11.13 for Hesstvedt's (1968) models of O and H_2O,
respectively.

Also, Shimazaki and Laird (1970) have extended Hunt's
(1966) non-equilibrium calculation for various neutral constituents in the mesosphere and lower thermosphere, by including the effects of molecular and eddy diffusion. Refer
to Figs. 11.14 and 11.15 for Shimazaki and Laird's (1970)
models of O and H_2O, respectively.

More recently, Hesstvedt (1970) has presented a meridional model of the oxygen-hydrogen atmosphere between 65 and
105 km. Refer to Fig. 11.16 for the vertical distribution
of H, H_2 and H_2O for 45° latitude in summer. Hesstvedt
(1970) considered vertical transport by eddies and by mean
motion, while horizontal transport was disregarded. He
concluded that the horizontal variations in the chemical
composition are relatively small and that the chemical composition of the upper mesosphere and lower thermosphere is
largely determined by photochemistry and vertical transport.

Anderson (1970) considered a mesospheric oxygen-hydrogen
model which includes diffusive transport of H_2O vapor in order

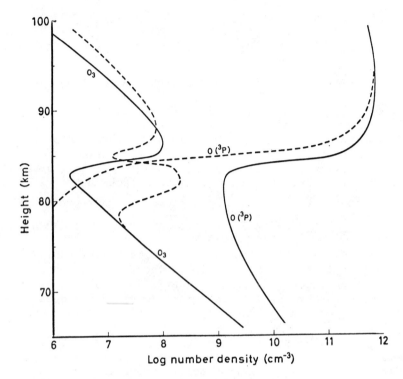

Fig. 11.12 Theoretical distribution of atomic oxygen, including eddy transport effects. Solid curves (day) and
dashed curves (night)   (Hesstvedt, 1968).

to interpret a rocket measurement of OH resonance fluorescence.   Anderson (1970) demonstrated that the upper limit
on the hydroxyl column density requires the eddy diffusion
coefficient to be less than 4×10^6 cm^2 sec^{-1} in the lower
mesosphere. Anderson's models for the concentration of H,
OH, HO_2 and H_2O in the mesosphere are shown in Fig. 11.17;
an H_2O vapor mixing ratio of 5×10^{-6} at 50 km and an eddy
diffusion coefficient of 1×10^6 cm^2 sec^{-1} were used in
calculating these models.

The important mechnisms for the mid-latitude, daytime
production of D-region electrons are believed to be: 2-8Å
solar X-rays in the upper D region; photoionization of O_2
($^1\Delta$) by solar UV in the middle and upper D region; photoionization of nitric oxide by solar Lyman alpha radiation in
the middle and upper D region; and galactic cosmic rays in
the lower D region below 70 km.

Figure 11.9 shows various nitric oxide concentration

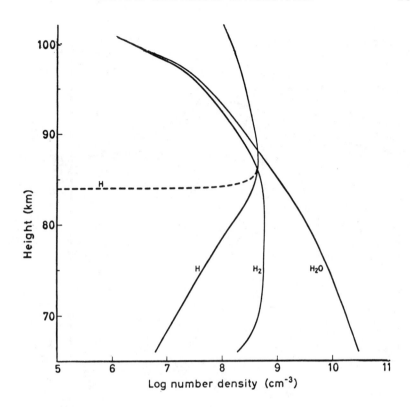

Fig. 11.13 Theoretical distribution of water vapor, including eddy transport effects. Solid curves (day) and
dashed curve (night)   (Hesstvedt, 1968).

profiles between 60 and 90 km. The nitric oxide profile by
Barth (1966) was based on a rocket experiment on November 17,
1963 at Wallops Island, Virginia, at 60° solar zenith angle.
Barth deduced a NO concentration at 85 km of 3.9×10^7 cm^{-3};
In Fig. 11.9 it was assumed that this value of NO concentration prevails at 85 km, and that the NO is completely mixed
with the atmosphere at all altitudes above and below this.
Pearce's (1969) NO profile is also shown in Fig. 11.9.
Pearce's measurements were obtained on December 12, 1967 at
Wallops Island, Virginia, when the solar zenith angle was
near 92° just after local ground sunset. Pearce showed that
above 75 km the dominant process controlling the nitric oxide
distribution is atmospheric mixing; and below 70 km, the
chemical reaction of nitric oxide with ozone was suggested as
the dominant process controlling the nitric oxide concentration.

    Another NO distribution shown in Fig. 11.9 is based on
theoretical work by Hesstvedt and Jansson (1969) who considered the effect of vertical eddy transport on the distri-

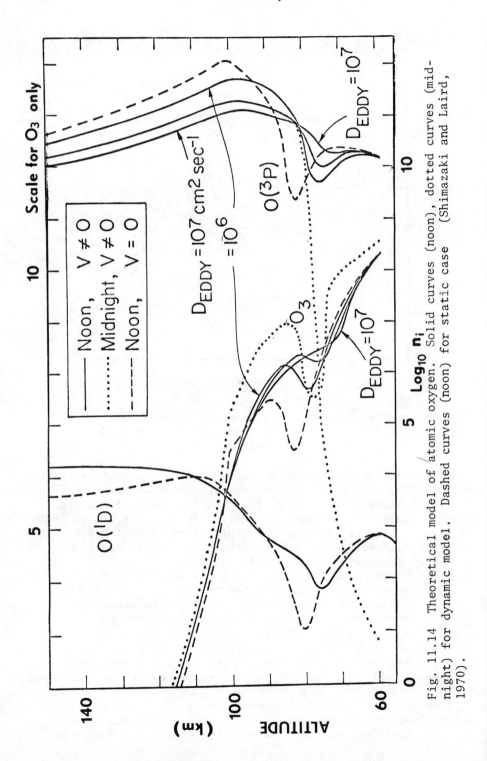

Fig. 11.14 Theoretical model of atomic oxygen. Solid curves (noon), dotted curves (midnight) for dynamic model. Dashed curves (noon) for static case (Shimazaki and Laird, 1970).

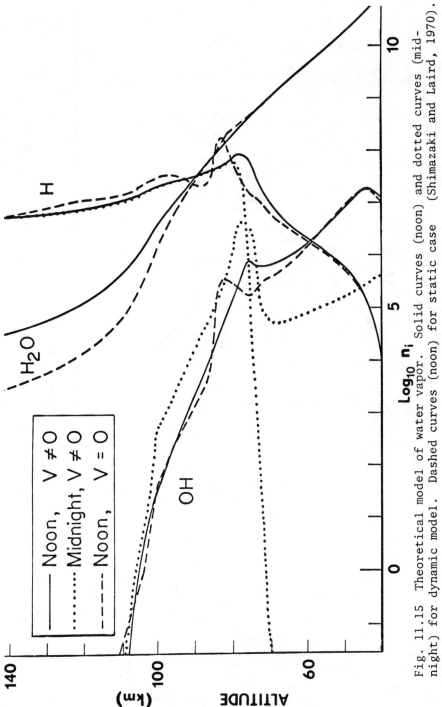

Fig. 11.15 Theoretical model of water vapor. Solid curves (noon) and dotted curves (mid-night) for dynamic model. Dashed curves (noon) for static case (Shimazaki and Laird, 1970).

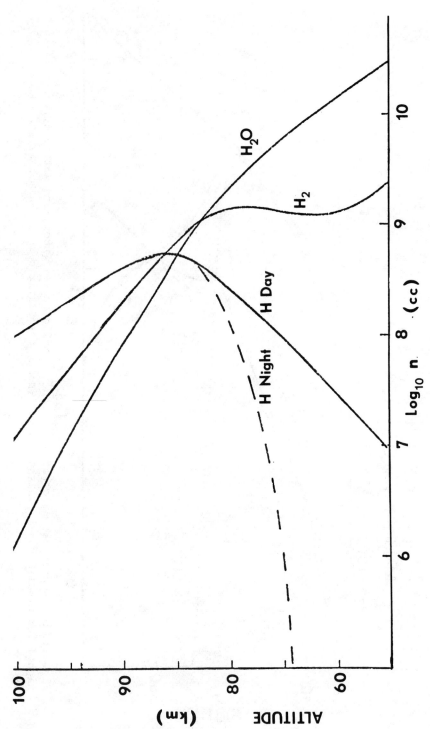

Fig. 11,16 Theoretical distributions of H, H_2 and H_2O (Hesstvedt, 1970).

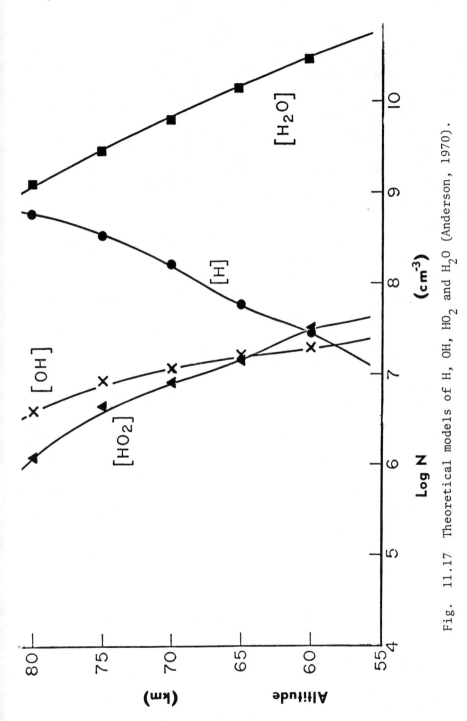

Fig. 11.17　Theoretical models of H, OH, HO$_2$ and H$_2$O (Anderson, 1970).

distribution of nitric oxide in the D region. They showed
that vertical eddy transport modifies the photochemical model
significantly and brings the theoretical results into much
closer agreement with the observation of Barth.

Rocket measurements of Lyman alpha flux, using ion
chambers, have been presented by Weeks (1967). He reported
solar Lyman alpha fluxes above the atmosphere of 4.6 and 4.9
ergs cm^{-2} sec^{-1} on December 15, 1965 and June 14, 1966, respectively. These fluxes correspond to photon fluxes of about
3×10^{11} photons cm^{-2} sec^{-1}, and this value has been assumed
for the calculation of electron production rates in this
paper.

Figure 11.18 presents the electron production rate distributions from the various ionization sources. The three
curves labelled H, B, and P refer to the electron production
rates calculated on the basis of Lyman alpha photoionization
of the nitric oxide distributions of Hesstvedt and Jansson,
Barth, and Pearce, respectively. The electron production
rates due to 2-8Å solar X-rays, the 33.7Å line of C VI, and
Lyman β are shown in Fig. 11.18; these production rates were
obtained from the work of Bourdeau, et al. (1966) who described the lower ionosphere at solar minimum.

The relevant equations used in the computations of the
H, B, and P curves in Fig. 11.18 are as follows:

$$P = \sigma_i (NO) \quad \{NO\} \ Q_\infty exp(-\tau) \qquad (11.10)$$

where P = electron production rate in cm^{-3} sec^{-1},
$\quad \sigma_i (NO)$ = ionization cross section of nitric oxide ($\lambda=1216Å$)
$\quad\quad\quad = 2 \times 10^{-18}$ cm^2,
$\quad \{NO\}$ = number density of nitric oxide in molecules cm^{-3},
$\quad Q_\infty$ = Lyman-alpha flux incident at top of the atmosphere=
$\quad\quad\quad 3 \times 10^{11}$ photons cm^{-2} sec^{-1}

and

$$\tau = \sigma_a (O_2) \int_z^\infty \{O_2\} \ sec \ \chi \ dz \qquad (11.11)$$

where τ = optical depth,
$\sigma_a (O_2)$ = absorption cross section of molecular oxygen ($\lambda=1216Å$)
$\quad\quad\quad = 1 \times 10^{-20}$ cm^2,

$\quad O_2$ = number density of molecular oxygen in molecules cm^{-3}
(CIRA, 1965),
$\quad \chi$ = solar zenith angle = 60°.

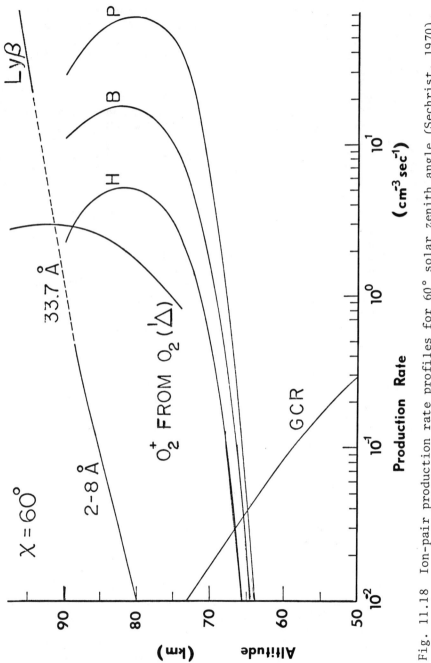

Fig. 11.18   Ion-pair production rate profiles for 60° solar zenith angle (Sechrist, 1970).

The curve in Fig. 11.18 labelled O_2^+ from $O_2(^1\Delta)$ is from the work of Hunten and McElroy (1968). They considered metastable $O_2(^1\Delta)$ as a major source of ions in the D region. In fact, Fig. 11.18 reveals that this source of ionization is competitive with Lyman alpha photoionization of nitric oxide, assuming the Hesstvedt and Jansson NO model.

Finally, Fig. 11.18 shows the ion-pair production rates due to galactic cosmic rays in the lower D region; the galactic cosmic rays produce ionization in the lower D region through a collisional ionization process, and the electron production rate profile shown is based on that given by Webber (1962) for 50° geomagnetic latitude and sunspot minimum conditions.

Hunten and McElroy (1968) pointed out that $O_2(a^1\Delta_g)$ can be ionized by the wavelength band 1027-1118Å, parts of which can penetrate below 70 km. Furthermore, they suggested that this source was able to explain the presence of O_2^+ ions in the D region, and they found that estimates of the O_2^+ production rates were comparable to those available for NO^+ from NO. O_2^+ ion production rates from $O_2(^1\Delta)$, for 60° solar zenith angle, were estimated to be: 1 cm^{-3} sec^{-1} at 75 km; 15 cm^{-3} sec^{-1} at 80 km; 2.5 cm^{-3} sec^{-1} at 85 km; and 3 cm^{-3} sec^{-1} at 90 km.

The foregoing ideas regarding the importance of $O_2(^1\Delta)$ have been questioned by Huffman, et al. (1971) who presented new results concerning the photoionization of $O_2(^1\Delta_g)$ in the ionosphere. Essentially, Huffman, et al. presented improved ion production rates based on new laboratory data including new $O_2(^1\Delta_g)$ photoionization cross sections, and more detailed curves in several ground-state O_2 windows. Also, recent solar flux measurements by Hinterreger and Hall reduce the continuum intensity by about a factor of 5. However, this reduction is partly compensated by including in the calculations the Si III multiplet, which is at the deepest O_2 window, and by the new ionization cross sections which are generally larger than the previously assumed values.

According to Huffman, et al., an important factor not considered previously is absorption by CO_2 which has a much larger absorption cross section than ground-state O_2 throughout this region. With the inclusion of CO_2 absorption in the O_2 windows, an ion-pair production rate of 1 cm^{-3} sec^{-1} is reached at approximately 80, 84, and 86 km for solar zenith angles of 0°, 45°, and 60°, respectively. Alternatively, production rates at 80 km are 1, 0.5 and 0.2 cm^{-3} sec^{-1} for solar zenith angles of 0°, 45° and 60°, respectively. The

rates from Hunten and McElroy (1968) for 60 zenith angle
were estimated to be 1 at 75 km, 1.5 at 80 km, 2.5 at 85 km,
and 3 cm^{-3} sec^{-1} at 90 km. Table 11.2 summarizes the foregoing results.

Table 11.2 Ion-pair production rates (cm^{-3} sec^{-1}) from
Hunten and McElroy (1968) and Huffman, Paulsen, Larrabee,
and Cairns (1971), at 80 km

χ	HM(1968)	HPLC(1971)
0°	3	1
60°	1.5	0.2

Thus, it is seen that the HM production rate is 3 times
larger than the HPLC rate for 0° zenith angle, and that the
HM rate is 7.5 times larger than the HPLC rate at 60° zenith
angle. One implication of these results is that $O_2(^1\Delta)$
photoionization may not be a competitive source of free
electrons in the daytime D region. A second implication of
the above results is that the hydrated positive ion reaction
scheme proposed by Ferguson and Fehsenfeld (1969) is in need
of revision; this is because the lower values of O_2^+ production from $O_2(^1\Delta)$ will decrease the theoretical 37^+ concentration significantly below the concentrations observed by Narcisi and Bailey (1965).

Electrons in the daytime D region are lost primarily by
the two processes of electron-ion (dissociative) recombination
and electron attachment. Some typical examples of dissociative recombination reactions are

$$NO^+ + e \xrightarrow{\alpha_{NO^+}} N + O$$

$$O_2^+ + e \xrightarrow{\alpha_{O_2^+}} O + O$$

$$XY^+ + e \xrightarrow{\alpha_{XY^+}} \text{neutral products}$$

where the dissociative recombination coefficients α_{NO^+} and
$\alpha_{O_2^+}$ have values of 5-10 x 10^{-7} cm^3 sec^{-1} at D-region temperatures. Assuming electron-ion recombination to be the dominant
electron loss process in the upper daytime D region, the total
electron loss rate would be given by

$$\Sigma L = \alpha_{NO^+}\{NO^+\}\{e\} + \alpha_{O_2^+}\{O_2^+\}\{e\} + \alpha_{XY^+}\{XY^+\}\{e\} + \ldots (11.12)$$

where the positive ion XY^+ represents an unknown (probably hydrated or water-clustered) positive ion, and α_{XY^+} denotes its recombination coefficient.

In the lower, daytime D region (below 75 km), negative ions become important, and a dominant electron loss process is believed to be 3-body attachment:

$$O_2 + O_2 + e \xrightarrow{k_a} O_2^- +.O_2$$

where the rate constant k_a has been measured in the laboratory and found to have the approximate value of 2.1×10^{-30} cm^6 sec^{-1}. It can be shown that, in the region where electron loss is dominated by electron attachment, the electron loss rate may be expressed as

$$L = f \, k_a \{O_2\}^2 \, \{e\} \qquad cm^{-3} \, sec^{-1} \qquad (11.13)$$

where $\{O_2\}$ and $\{e\}$ are the molecular oxygen and electron concentrations, respectively. In the region of interest, $f \ll 1$ and the attachment coefficient may be defined as $\beta \equiv f \, k_a \{O_2\}^2 \, sec^{-1}$. Since f is always less than or equal to unity, we can write

$$\beta_{max} \equiv k_a \{O_2\}^2 \, sec^{-1}$$

The continuity equation for electrons, assuming quasi-equilibrium conditions, is

$$\frac{d\{e\}}{dt} = \Sigma P - \Sigma L \overset{\sim}{=} 0 \qquad (11.14)$$

where ΣP and ΣL represent the sum of the electron production and electron loss rates, respectively.

The electron production rate at 80 km, based on the values given in Fig. 11.18, is about 10 electrons cm^{-3} sec^{-1}. For quasi-equilibrium conditions,

$$L \overset{\sim}{=} P \overset{\sim}{=} 10 \, cm^{-3} \, sec^{-1}$$

If electron-ion recombination is dominant at 80 km, then

$$P = \alpha \{N^+\}\{e\} = \alpha \{e\}^2 \qquad (11.15)$$

If $\{e\}_{80} \overset{\sim}{=} 10^3 \, cm^{-3}$, as shown by numerous rocket measurements (Mechtly and Smith, 1968), then

$$\alpha \overset{\sim}{=} \frac{P}{\{e\}^2} \overset{\sim}{=} \frac{10}{10^6} \overset{\sim}{=} 10^{-5} \, cm^3 \, sec^{-1}$$

One implication of this relatively high value for α is that neither NO^+ nor O_2^+ can be responsible for electron-ion recombination at 80 km, because of their relatively low ($\sim 10^{-6}$ $cm^3\ sec^{-1}$) values of recombination coefficient.

If electron attachment is assumed to be a dominant electron loss process at 80 km, then we can write

$$P = \beta\{e\}$$

For values of P and $\{e\}$ equal to 10 $cm^{-3}\ sec^{-1}$ and $10^3\ cm^{-3}$, respectively, we find

$$\beta = \frac{P}{\{e\}} = \frac{10}{10^3} = 10^{-2}\ sec^{-1}$$

But a value of $10^{-2}\ sec^{-1}$ is comparable to the value of β_{max} at 80 km.

Therefore, we are forced to conclude that the dominant electron loss process at 80 km is either recombination with an unknown positive ion—or attachment to an unknown neutral specie other than O_2. We prefer the former possibility because of the known presence of hydrated (water-clustered) positive ions (Narcisi and Bailey, 1965). Also, it would be very difficult to imagine an alternative electron attachment process that would be faster than 3-body attachment with molecular oxygen.

One of the best ways to determine the effective electron-ion recombination coefficient in the D region is to measure the electron density profile by means of two rockets between second and third contact during a solar eclipse. A rocket launched a few seconds after second contact and another rocket launched about 30-60 seconds afterwards would permit one to estimate the effective attachment rate of electrons in the lower D region below 75 km. A third rocket launched approximately 2 minutes following the first would permit an estimation of the effective electron-ion recombination coefficients in the middle and upper D regions.

For solar eclipse data, the simplest method of estimating the value of the electron-ion recombination coefficient α_e is to consider conditions when the electron production rate is near zero so that the continuity equation for electrons becomes

$$\frac{d\{e\}}{dt} = -\alpha_e\{XY^+\}\{e\} \qquad (11.16)$$

where $\{XY^+\}$ is the concentration of the dominant positive ion. In the middle and upper parts of the D region, above approximately 75 km, it is believed that the negative ion density is small compared to the electron density, and therefore it is valid to assume that the electron density is equal to the positive ion density (Reid, 1970a; Sechrist, 1970). Therefore, equation (11.16) becomes

$$\frac{d\{e\}}{dt} = -\alpha_e \{e\}^2 \tag{11.17}$$

The solution of Eq. (11.17) for the electron density is

$$\{e\} = \frac{\{e\}_0}{1+\alpha_e\{e\}_0\Delta t} \tag{11.18}$$

where $\{e\}_0$ is the electron concentration immediately following second contact and $\{e\}$ is the electron concentration at a time Δt seconds afterwards. The expression for α_e, the effective recombination coefficient, becomes

$$\alpha_e = \frac{\{e\}_0-\{e\}}{\{e\}_0\{e\}\Delta t} \tag{11.19}$$

Equation (11.19) may be applied to the results of Mechtly, et al. (1969) who measured lower ionosphere electron densities during the solar eclipse of November 12, 1966 in Brazil; these appear in Fig. 11.19. Rocket No. 2 was launched at 14^h08^m GMT when the solar zenith angle was near 20°. On ascent, rocket No. 2 entered totality at an altitude of about 82 km; at 70 km, only one-tenth percent of the solar disk area was visible at the position of this rocket. Rocket No. 3 was launched at approximately 14^h10^m GMT, and its trajectory was in totality from the ground up to an altitude of about 85 km. Between 85 and 100 km, rocket No. 3 was in a position where less than one-tenth percent of the solar disk area was visible. Because rockets No. 2 and 3 were in totality around an altitude of 82 km, it is possible to use Eq. (11.19) to estimate the effective recombination coefficient α_e in this altitude range. Equation (11.19) was solved, using the rocket electron densities in Fig. 11.19, and the results are presented in Fig. 11.20.

Figure 11.20 is a plot of the effective recombination coefficient in the altitude range between 75 and 90 km, and it is apparent that at least three different electron loss processes must be considered in the D region. In the upper D region, α_e lies between 10^{-6} and 10^{-5} cm^3 sec^{-1}; between 78 and about 86 km, α_e appears to be nearly constant with a value around 4×10^{-5} cm^3 sec^{-1}; and below 78 km, it is apparent that α_e increases sharply from 4×10^{-5} to over

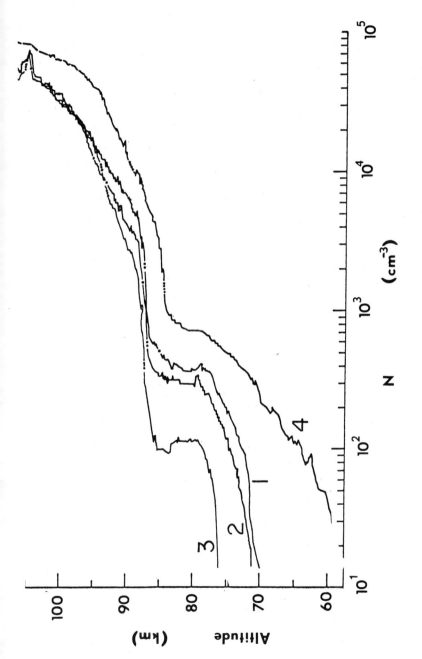

Fig. 11.19 Rocket electron density profiles during the solar eclipse of November 12, 1966 (Mechtly, et al., 1969).

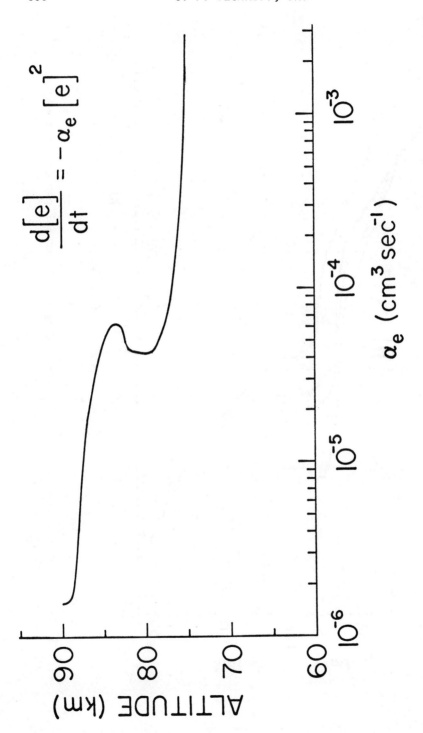

Fig. 11.20   Effective recombination coefficient deduced from eclipse data (Sechrist, 1970).

10^{-3} cm^3 sec^{-1}. Because of the improbability that electron-
ion recombination is an important electron loss process below
about 75 km, it appears that electron attachment is the dominant electron loss process below about 75 km, and therefore
Equation (11.16) should be modified for the region below about
75 km. Equation (11.20) is therefore thought to be applicable
to the region below 75 km.

$$\frac{d\{e\}}{dt} = - \beta_e \{e\} \qquad (11.20)$$

The solution of Equation (11.20) is

$$\{e\} = \{e\}_o \exp(-\beta_e \Delta t) \qquad (11.21)$$

where $\{e\}_o$ and $\{e\}$ are the electron concentrations measured
by the second and third rockets above, respectively. β_e
is the effective electron attachment frequency.

It may be concluded from Fig.11.20 that electron loss in
the upper D region above 85 km is due primarily to dissociative recombination of O_2^+ and NO^+ ions; between about 75 and
85 km, the dominant electron loss process is probably electron-ion recombination of 37^+; and that below 75 km electron
attachment appears to become the dominant electron loss process.

Reid (1970a) has also deduced profiles of the effective
and dissociative recombination coefficients in the daytime D
region, using a different approach. Essentially, Reid (1970a)
calculates α and α_d on the basis of measured electron densities and calculated electron production rates. He also
concludes that α at 80 km must have the value 4-5 x 10^{-5}
cm^3 sec^{-1} if the NO concentration of Barth (1966) is accepted.

The abrupt change in recombination coefficient around 82
km, noted by Reid (1970a) and Sechrist (1970), is possibly
due to the abrupt decrease in the relative abundance of hydrated positive ions which should have recombination coefficients significantly greater than those for NO^+ and O_2^+ ions.
Based on the positive-ion composition results of Narcisi and
Bailey (1965), as shown in Fig. 11.21, it seems reasonable
to suggest that the quasi-equilibrium electron density in the
upper D region may be expressed as

$$\{e\} = \frac{\Sigma p}{\alpha_{NO^+}\{NO^+\}+\alpha_{O_2^+}\{O_2^+\}+\alpha_{37^+}\{37^+\}} \qquad (11.22)$$

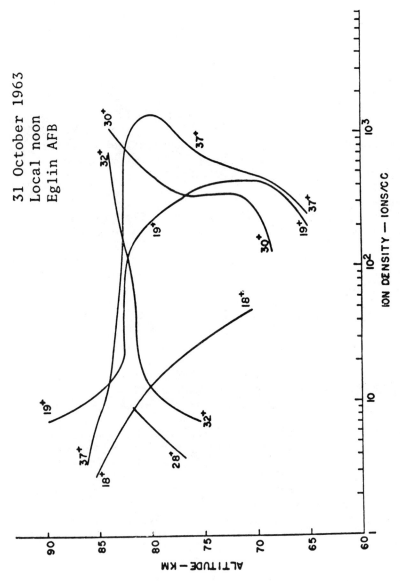

Fig. 11.21 Positive ion composition profiles (Narcisi and Bailey, 1965).

Below the observed steep gradient in the electron density profile around 82 km, the term $\alpha_{37^+}\{37^+\}$ dominates in the denom-
inator of Equation (11.22). Above the electron density ledge
at 82 km, the hydrated positive ion concentration drops
markedly and the electron density is given simply by

$$\{e\} = \frac{\Sigma P}{\alpha_{NO^+}\{NO^+\}+\alpha_{O_2^+}\{O_2^+\}} \qquad (11.23)$$

Fehsenfeld and Ferguson (1969) proposed a water vapor
phase reaction scheme. Therefore, an approximate expression
for the 37^+ concentration is

$$\{H_5O_2^+\} \approx \frac{P(O_2^+)}{k_9\{H_2O\}\{M\}} \qquad (11.24)$$

It is important to point out that the calculations of
Ferguson and Fehsenfeld (1969) were carried out using the
very large NO concentrations of Pearce (1969). Lowering the
NO concentrations to the values of Barth (1966) would nearly
double the calculated $H_5O_2^+$ concentration. However, according
to Reid (1970b) even Barth's NO concentrations may be too
large.

Furthermore, because of the fact that Hunten and McElroy
(1968) did not take into account mesospheric CO_2 absorption of
the solar UV band that can ionize $O_2(^1\Delta)$, it appears that the
ionization production rates of O_2^+ from $O_2(^1\Delta)$ below 80 km
are considerably less than the values used by Ferguson and
Fehsenfeld (1969).

Finally, it must be realized that because of the uncertainties in the NO and H_2O concentrations, and the lower O_2^+
production rates, it may be necessary to revise the semi-
quantitative estimates of the 37^+ concentrations.

A further complication with the Ferguson and Fehsenfeld
(1969) scheme involved the steep gradient in 37^+ concentration around 82 km. They attributed this ledge to a rapid
drop in the H_2O vapor concentration, but unfortunately all of
the H_2O models (which include vertical eddy transport effects)
do not reveal an abrupt decrease in water vapor concentration.
Apparently this discrepancy has been resolved by Ferguson
and Fehsenfeld (1970) who observed that the reaction

$$O_4^+ + O \rightarrow O_2^+ + O_3$$

occurs in the laboratory. Thus, a possible implication of
this is that the steep ledge in both the 37^+ and electron

density distributions is due to an abrupt decrease in the
atomic oxygen concentration near 82 km.

It has been observed qualitatively by Ferguson and
Fehsenfeld (1970) that the reaction

$$O_4^+ + O \xrightarrow{k_{10}} O_2^+ + O_3 \qquad (11.25)$$

occurs. This will have an effect on the water cluster ion
production scheme proposed by Ferguson and Fehsenfeld (1969),
which was given by the reaction sequence

$$O_2^+ + O_2 + M \rightarrow O_4^+ + M \qquad (11.26)$$

$$O_4^+ + H_2O \rightarrow O_2^+ \cdot H_2O + O_2 \qquad (11.27)$$

$$O_2^+ \cdot H_2O + H_2O \rightarrow H_3O + OH + O_2 \qquad (11.28)$$

$$\rightarrow H_3O^+ \cdot OH + O_2 \qquad (11.29)$$

$$H_3O^+ \cdot OH + H_2O \rightarrow H_3O^+ \cdot H_2O + OH \qquad (11.30)$$

Reaction (11.25) increases the D-region O_2^+ concentrations, and, with a likely rate constant, leads to the sharp
decrease in $H_5O_2^+$ concentration above 82 km that has been
observed repeatedly by Narcisi, et al. By incorporating
reaction (11.25) into their earlier hydrated positive ion reaction scheme, Ferguson and Fehsenfeld (1970) recalculated
the positive ion profiles, using a range of values for the
rate constant k_{10}. Their calculations yielded upper and
lower limit estimates for the H_2O mixing ratio in the 80 km
height region. Also, they were able to derive a lower limit
on the eddy diffusion coefficient.

Because reaction (11.25) is fast, it interrupts the O_2^+
$\rightarrow H_5O_2^+$ reaction sequence. Although the rate constant k_{10}
has not been determined yet, it appears likely that $k_{10} \approx$
10^{-10} cm^3 sec^{-1}. One effect of including reaction (11.25)
is to produce the steep ledge in the $H_5O_2^+$ profile that has
been observed repeatedly. If the Hunten and McElroy (1968)
O_2^+ production rates and 1 ppm H_2O are assumed, Ferguson (1970b)
found that the fit between calculated and observed $H_5O_2^+$ profiles was quite good with $k_{10} \approx 10^{-10}$ cm^3 sec^{-1}. (However,
it should be remembered that Huffman, et al. pointed out
that the Hunten and McElroy O_2^+ production rates were probably
too high by a factor of roughly 10 for 60° zenith angle.)
Thus, it appears that Ferguson's (1970b) calculated $H_5O_2^+$
profiles are no longer in close agreement with the observed
ones. However, it is known that the Narcisi and Bailey

(1965) ion densities are too large because of normalization
to probe measurements, and it is believed that the lower
altitude densities are exaggerated by the method of data
reduction and by breakup of heavier water cluster ions in the
rocket mass spectrometer sampling process.

Recapitulating, the lower O_2^+ production rates of Huffman,
et al. (1971) now lead to a severe problem in producing a
significant $H_5O_2^+$ concentration at the lower D-region altitudes.

The reaction scheme proposed by Fehsenfeld and Ferguson
(1969) and Ferguson and Fehsenfeld (1969) omits the NO^+ problem (Reid, 1970b). At middle latitudes (where energetic
particle effects are relatively unimportant) there is a need
for a fast reaction sequence to convert NO^+ ions to water
cluster ions, since NO^+ (from atmospheric measurements of
both NO concentration and Lyman α flux) appears to be produced at a substantially faster rate than O_2^+; yet NO^+ is not
observed as the predominant ion below 80 km.

The lower O_2^+ production rates calculated by Huffman,
et al. (1971) now lead to a severe problem in producing a
significant $H_5O_2^+$ concentration at the lower altitudes (h
$\lesssim$ 70 km). In view of this development, it is necessary to
re-examine the role of NO^+ ions in leading to water cluster
ions (Reid, 1970b). Ferguson (1970b) believes that NO^+
leads to $H_7O_3^+$ (55$^+$) production and not to $H_5O_2^+$ (37$^+$) production. Thus, it becomes imperative to know what is the true
ambient positive ion composition below $\sim$ 75 km before further
progress can be made in D-region positive ion chemistry.
In particular, we must know to what extent the 37$^+$ concentration (observed as low as 65 km) is representative of the
positive ion composition, and to what extent it is due to
fragmentation of ambient 55$^+$ ions in the rocket mass spectrometer sampling process.

A major problem of D-region ion chemistry is the matter
of NO^+ production and loss. If current estimates of NO^+ production rates (based on Barth's or Meira's NO profiles) are
accepted, it is necessary to require a faster NO^+ loss rate
than is presently known.

Assuming the water vapor reaction scheme (involving NO^+
and its hydrates) proposed by Fehsenfeld and Ferguson (1969),
it is important to note that the smallest hydrated water
cluster ion produced is $H_7O_3^+$ (55$^+$); whereas the major water
cluster ion observed in the D-region is $H_5O_2^+$ (37$^+$). However,
the extent to which the fragmentation process ($H_7O_3^+$ + 1 eV $\rightarrow$
$H_5O_2^+$ + H_2O) occurs in rocket sampling is unknown. According

to Narcisi (1970a), rocket mass spectrometer measurements have shown evidence of thermodynamic decomposition of positive water cluster ions in the shock-heated gas and also of electric field-induced collisional fragmentation of both positive and negative water cluster ions. Furthermore, Goldberg and Aikin (1970) conducted rocket-borne positive ion mass spectrometric measurements at Thumba, India and found that heavy water cluster ions were present on rocket descent but not on rocket ascent. They surmised that the shock-heated region adjacent to the sampling orifice fragmented the heavy cluster ions on rocket ascent, and that the wake sampling on rocket descent resulted in less fragmentation.

Laboratory workers (Ferguson, 1970b) are attempting to determine the major NO^+ loss processes in the D region, and preliminary indications are that the $NO^+ \cdot O_2$ and $NO^+ \cdot N_2$ cluster ions are too weakly bound to be significant for NO^+ loss. The main loss process thus far discovered is the association with CO_2, namely

$$NO^+ + CO_2 + N_2 \rightarrow NO^+ \cdot CO_2 + N_2$$

The relatively large CO_2 abundance makes the above reaction faster than dissociative recombination ($NO^+ + e \rightarrow N + O$) below 80 km. It is believed that the reaction

$$NO^+ \cdot CO_2 + H_2O \rightarrow NO^+ \cdot H_2O + CO_2$$

is very fast ($k \sim 10^{-9}$ cm^3 sec^{-1}), based on qualitative observations and from a generalization from observations on many similar reactions (Adams, et al., 1970). Also, the $NO^+ + CO_2 + N_2$ reaction followed by the $NO^+ \cdot CO_2 + H_2O$ reaction is probably faster than direct hydration because $\{CO_2\} \sim 10^2\{H_2O\}$, while the rate constant for $NO^+ + CO_2 + N_2$ is greater than 10^{-2} times the rate constant for $NO^+ + H_2O + N_2$.

Recapitulating, the NO profiles deduced by Barth and Meira have indicated that NO^+ ions are produced directly in the D region in greater quantities than O_2^+ ions, in addition to being formed indirectly by charge exchange with O_2^+. While NO^+ can give rise to the water-cluster ions (Fehsenfeld and Ferguson, 1969), the reaction chain is not nearly as efficient or rapid as that originating with O_2^+ ions, and it appears qualitatively as though the dominant positive ions in the vicinity of 80 km are likely to be NO^+ and its first two hydrates $NO^+ \cdot H_2O$ (48^+) and $NO^+ \cdot 2(H_2O)$ (66^+). The electron-ion recombination coefficients of the NO^+ hydrates

are probably comparable with those of the H_3O^+ hydrates,
and their presence in large quantities would be compatible
with the observed electron densities. However, dominance
of the NO^+ hydrates would certainly be contrary to the mass-
spectrometer results.

In order to examine the above problem, Reid (1970b)
carried out computations for a reaction scheme involving
16 species of positive ions, and assuming an altitude of 80
km which lies below the normal location of the electron density ledge. Reid found that the estimates of ion production
rate were not in agreement with observations as follows:
First, they predicted that the dominant ion species should be
NO^+ and its hydrates rather than the water-cluster ions that
have been observed consistently to be dominant at 80 km;
secondly, they predicted electron densities that were much
larger than the observations allowed. Hence, it is now
difficult to explain this observed dominance of the water-
cluster ions in view of the large production rate of NO^+
ions which apparently can be converted to the water-cluster
ions very slowly.

Reid (1970b) demonstrated that a problem exists with
both the O_2^+ and NO^+ production rates, in that the estimated
values are individually too large to account for the observed
electron densities even if they operate alone. However,
he pointed out that the problem is less severe now because
Huffman, et al. (1971) have shown that the O_2^+ production rates
of Hunten and McElroy (1968) should be revised downwards by
about an order of magnitude. Unfortunately, even with the
smaller values of O_2^+ production rate, Reid computed a predominantly NO^+- derived composition around 80 km, assuming
a reduction of NO^+ production rate by about a factor of 4
from that predicted by the NO observations.

Reid (1970b) surmised that the only way in which we
can achieve simultaneously electron densities that are comparable with those observed, and an ion composition that is
predominantly formed of water-cluster ions, is to increase
O_2^+ production to a value between those predicted by Hunten
and McElroy, and Huffman, et al., and to reduce NO^+ production to a value well below that based on Barth's or Meira's
NO densities.

Reid (1970b) concluded that present estimates of NO^+
production rate are too high by about an order of magnitude
to explain either the electron density or the relative positive-ion composition at 80 km altitude. Essentially, this
conclusion is based on the fact that there is no known

efficient reaction operating to remove NO^+ ions and produce
water-cluster ions instead, analogous to the clustering suffered by O_2^+ with O_2. One is thus forced to the conclusion
that present estimates of the NO concentration may be suspect.

It should be noted that the Ferguson and Fehsenfeld
(1969) reaction scheme for hydrated positive ions is more
directly applicable to disturbed D-region conditions. It
seems that there is much less of a problem in explaining the
origin of water cluster ions from energetic ionization sources
which can ionize any atmospheric constituent more or less
indiscriminately (Narcisi, 1970b). Essentially, any of the
major constituent ions O_2^+, N_2^+, A^+, etc., will (immediately
following initial ionization) rapidly convert to O_2^+ by
charge transfer in less than 10^{-4} seconds as follows:

$$M^+ + O_2 \overset{k}{\longrightarrow} O_2^+ + M$$

where $k \approx 10^{-10}$ cm^3 sec^{-1} and $\{O_2\} > 10^{14}$ cm^{-3} below 85 km.
The O_2^+ produced then initiates the Ferguson and Fehsenfeld
reaction sequence starting with the reaction $O_2^+ + O_2 + O_2 \rightarrow$
$O_4^+ + O_2$. Also, during moderately disturbed periods the
ionization created by the usual solar radiation represents
only a small portion of the total ionization and can be
neglected. Thus, if the Ferguson and Fehsenfeld sequence
$(O_2^+ \rightarrow O_4^+ \rightarrow O_2^+ \cdot H_2O \rightarrow H^+ \cdot n(H_2O))$ is operative, O_2^+ is converted to water-cluster ions in a matter of seconds at 80 km
and much shorter times at lower altitudes.

Under normal, undisturbed ionospheric conditions at Ft.
Churchill near sunset and during the night, Narcisi, et al.
(1970) observed positive water-cluster ions predominate up to
86 km. However, during the November 1969 polar cap absorption
event at Ft. Churchill, positive ion composition measurements
performed near midday and midnight revealed that water cluster
ions during the day, night and sunset were either absent or
in small concentrations above $\sim$ 75 km but predominant below
70 km; and the major positive ions above 75 km were NO^+ and
O_2^+.

Because very few water-cluster ions were observed in the
region 75-86 km, and NO^+ and O_2^+ were the major ions present,
Narcisi, et al. (1970) surmised that the $O_2^+ \rightarrow O_4^+ \rightarrow O_2^+ \cdot H_2O \rightarrow$
$H^+ \cdot n(H_2O)$ reaction sequence had to be broken in such a way
as to regenerate O_2^+ ions. Narcisi, et al. (1970) suggested
that the O_2^+ ions then apparently produced the major ion NO^+
by $O_2^+ + NO \rightarrow NO^+ + O_2$ charge transfer. Alternatively, they
further suggested that the water-cluster ions have very
large electron-ion recombination coefficients, and because of

the large electron densities ($\sim 2 \times 10^4$ cm^{-3}) these ions are quickly destroyed. However, the preliminary laboratory measurements (Biondi, 1970) indicate that the water-cluster ion recombination coefficients are not large enough to support the alternative suggestion above.

The reactions suggested to break the $O_2^+ \rightarrow O_4^+ \rightarrow O_2^+ \cdot H_2O \rightarrow H^+ \cdot n(H_2O)$ reaction sequence were

$$O_4^+ + O \rightarrow O_2^+ + O_3$$

$$O_2^+ \cdot H_2O + O_2(^1\Delta) \rightarrow O_2^+ + \text{products}.$$

Large amounts of $O_2(^1\Delta)$ were produced and observed during the PCA event between 70 and 90 km. Thus, Narcisi, et al. (1970) suggested that O and $O_2(^1\Delta)$ are responsible for precluding the production of water-cluster ions and thereby increasing the duration of the D-region ionization because both NO$^+$ and O_2^+ ions have electron-ion recombination coefficients smaller than those for the water-cluster ions.

In conclusion, it should be mentioned that the Max Planck Group conducted a rocket-borne positive ion mass spectrometer experiment during a strong aurora at Andøya, Norway in April 1969. The results were somewhat similar to the PCA results of Narcisi, et al. which showed that the water cluster ions were even more diminished above 75 km.

Electron densities in the D region may be enhanced by the indirect effects of energetic particle precipitation. That is, even though ion-pair production rate may be increased due to particle precipitation, it appears that positive ion composition may be drastically altered. For example, during auroral and PCA absorption events the major ions observed in the mid and upper D region were NO$^+$ and O_2^+ instead of the hydrated species. Since NO$^+$ and O_2^+ have recombination coefficients $\sim 10^{-6}$ cm^3 sec^{-1} while hydrated positive ions may have much higher values of recombination coefficient, it seems likely that electron density enhancements may be caused by changes from hydrated positive ion species to non-hydrated simple molecular species such as NO$^+$ and O_2^+.

Figure 11.6 shows rocket electron density profiles measured on normal and anomalous winter days at Wallops Island, Virginia (Sechrist, et al., 1969). It has been demonstrated that the bulk of anomalous winter absorption of MF and HF radio waves occurs in the 80 to 90 km height range. Therefore, the problem of the winter anomaly is actually a problem involving the positive-ion chemistry of the region between

80 and 90 km, approximately. Hence, variability of electron
density in this height interval may be ascribed to variability of one or more of the following minor neutral constituents:
NO, $O_2(^1\Delta)$, O and H_2O. Furthermore, it is important to consider the effects of energetic particles on the positive-ion
composition, because rocket mass spectrometric measurements
during solar proton and auroral precipitation events have
revealed that hydrated positive ions are present only in
the lower D region.

The foregoing results suggest that the anomalous electron
density enhancements above 80 km in winter at middle latitudes
may be caused by an abrupt change in the electron loss process above about 80 km. That is, both atomic oxygen and
energetic particles can influence the positive-ion composition in the upper D region and determine the altitude at
which the positive-ion composition switches abruptly from
water-clustered ions (having high recombination coefficients)
to simple molecular ions like NO^+ and O_2^+ (which have relatively lower recombination coefficients). Thus, variability
in upper D-region electron densities may be ascribed to variability in the distribution of minor neutral constituents,
and this in turn is influenced strongly by various transport
processes in the neutral atmosphere.

11.4   Theories of Neutral Atmosphere-Lower Ionosphere
Interactions

One of the most *perplexing* problems of the mid-latitude
D region of the ionosphere is the origin of the temporal and
spatial variability of electron density during the winter
months. This variability has been known to exist since the
1930's and has been manifested in high-frequency (HF) radio
wave absorption measurements, low-frequency (LF) and very-
low-frequency (VLF) phase and amplitude measurements, rocket
measurements of D-region electron concentration, and ground-
based measurements of electron concentration using the wave
interaction and partial reflection techniques. Without doubt,
winter variability of D-region electron densities (including
the winter anomaly) is one of the most *convincing manifesta-
tions* of interaction between the neutral atmosphere and the
lower ionosphere.

Thus far, *meteorological* and *energetic particle* sources
have been proposed to explain the winter variability of D-
region electron concentration. Unfortunately, it has not
been possible to ascertain the relative importance of these
two proposed sources. Changes in D-region electron density
can be realized in several ways, assuming that solar ionizing

and dissociating radiations remain constant. For example, the electron production rate in the middle D region is dominated by the photoionization of nitric oxide by solar Lyman alpha (1216Å) radiation. Any changes in the concentration profile of nitric oxide would certainly affect the electron concentration.

Variations in electron density can occur through changes in the electron-ion recombination coefficient also. For example, Reid (1970a) and Sechrist (1970) have suggested that there is an abrupt change in the recombination coefficient near 82 km because of the sudden decrease in hydrated positive ions near that altitude. Reid (1970a) and Sechrist (1970) show that a recombination coefficient of $\sim 10^{-5}$ cm^3 sec^{-1} is required below 82 km, while above that a value near 10^{-6} cm^3 sec^{-1} is necessary. One implication of this is that changes in the water vapor concentration profile can modify the hydrated positive ion concentrations, and this in turn would alter the height dependence of the electron-ion recombination coefficient. Therefore, changes in the H_2O concentration profile could seriously alter the electron concentration.

At lower altitudes in the D region, negative ion chemistry becomes important, and electron loss processes are more complex. Some of the important minor, neutral constituents which influence the electron density include atomic oxygen, ozone, nitric oxide, carbon dioxide, water vapor, and others. The point is that dynamically induced or chemically induced changes in these minor, neutral constituents can produce electron density variations. At present, it is believed that the thermal structure and the altitude distributions of minor, neutral constituents can be modified by transport processes in the neutral mesosphere and lower thermosphere. These transport processes include: horizontal advection, subsidence, mean vertical motions induced by large-scale meridional circulation cells and planetary waves and eddy diffusion induced by effects of gravity waves. Hence, there is reason to believe that neutral atmosphere dynamics plays a significant role in the creation and maintenance of electron density variations in the winter D region.

Alternatively, Maehlum (1967) has suggested that energetic electrons with energies greater than 40 keV play an important role in the mid-latitude, winter daytime D region. He pointed out that the flux of >40 keV electrons, measured by satellites, was sufficient to perturb the winter daytime electron density in the D region; and he stated that precipitating electrons may be at least in part responsible for the phenomenon of anomalous D-region radio wave absorption. (The D-region winter anomaly may be considered to be an extreme case of winter variability.) More recent studies by Belrose and Thomas (1968) and Bourne and Hewitt (1968) have presented convincing evidence for precipitating particle influences in the D region. Interestingly, Lauter and Knuth (1967) considered precipitation of high energy particles into the upper atmosphere at medium latitudes after magnetic storms, and their summary of the features of the after effect of strong magnetic storms in low-frequency radio wave absorption indicated the possibility of patrolling high energy particle flux conditions in the upper atmosphere below 90 km.

Recently, Manson and Merry (1970) reviewed the two explanations of the D-region winter anomaly of high-frequency radio wave absorption; these are (a) changes in the neutral and ionized components of the atmosphere because of meteorological influences; and (b) enhancements in, or the introduction of, an ionizing corpuscular flux of precipitated particles. They concluded that (a) was more important than (b), and that (a) and (b) are linked because seasonal changes in the neutral density below 90 km are likely to be related to meteorological influences, and yet are of importance in establishing ionization rates due to particle influx events.

Particularly during the winter months, the electron density in the ionospheric D region shows considerable variation from day to day, and during a single day. This variability is a matter of considerable interest at present. Ground-based soundings show the existence of correlations between this variability and properties of the stratosphere (Bowhill, 1969a, b), and rocket measurements (Mechtly and Smith, 1968; Sechrist, et al., 1969) have demonstrated that the shape of the electron density profile can be quite different on days when the ionization is enhanced (the so-called *winter anomalous days*).

Changes in concentrations of minor neutral constituents have been proposed by many ionospheric workers to account for variations of mid-latitude D-region electron densities in winter. According to Nicolet (1955) atmospheric motions will affect the production and downward transport of nitric oxide (NO), and increases in the mesospheric electron densities are to be expected even when the Sun is quiet. Nicolet (1962) suggested a possible explanation for the winter anomaly in terms of an increase of the NO concentration at the mesopause level. According to Nicolet (1962, 1965), the photochemical concentration of NO depends on the variation of that of atomic nitrogen (N), and values of (NO) should be between

1×10^{-10} and 2×10^{-10} of the total concentration, assuming
photochemical equilibrium conditions. However, Nicolet
(1962) correctly pointed out that any disturbance (in transport processes) modifies the photochemical distribution of
NO, and thus the conditions for the vertical distribution of
NO in the mesosphere must be associated with mesospheric
motions leading to mixing. Nicolet (1962) stated that variations of (NO) at and below the mesopause will lead to variations of the electron density associated with atmospheric
motions and not with variations of solar Lyman-α radiation.

Mawdsley (1961) ascribed the anomalous increases in
winter absorption of radio waves to increases in the abundance
of nitric oxide relative to molecular oxygen; and he suggested
that the large air-density increases observed in winter at
Fort Churchill by Jones, et al. (1959) may reveal vertical
air movements that bring about the much larger enhancement in
NO concentration suggested by Nicolet (1955) and give rise to
increased daytime radio wave absorption.

Sechrist (1967) suggested that the NO concentration in
the D region may be enhanced on anomalous winter days because
of a warming of the mesopause region. Essentially, the photochemical equilibrium concentration of NO is highly temperature
sensitive, and Sechrist (1967) surmised that moderate increases in mesopause temperature would cause substantial increases in the NO concentration. The larger NO concentration
would then lead to a larger production rate for electrons and
thus an increased electron density. Sechrist (1967) did not
specify the source of the mesopause warming but presumably
it could be produced by a downward transport, or subsidence,
which could produce chemical heating through the mechanism of
atomic oxygen recombination as discussed by Kellogg (1961)
and Young and Epstein (1962).

Gregory (1965) observed that during a winter period at
Christchurch, New Zealand (43°S) isopleths of electron density descended and ascended by about 10 km, while stratospheric temperatures rose and fell by 10°C. Gregory ascribed
both effects to vertical motion in the neutral atmosphere,
and suggested that atmospheric circulation changes may cause
a redistribution of minor neutral constituents in the lower
ionosphere. Essentially, Gregory surmised that disturbances
of the winter westerly circulation in the mesosphere cause a
redistribution of ionizable photochemical constituents, particularly by virtue of vertical motion and possibly through
turbulent transport.

Geisler and Dickinson (1968) pointed out that both

planetary waves and a mean meridional cell will be important for the aeronomic problem of nitric oxide in the winter D region. Furthermore, they have suggested that the time dependent behavior of the winter anomaly may be related to nitric oxide transport by transient planetary waves.

Gregory and Manson (1969) have shown that the seasonal circulation changes and the perturbations produced by waves can provide a qualitative explanation of the observed seasonal variability of electron densities. The most direct evidence for the effect of wave motions upon electron densities, though limited to correlations between stratospheric and mesospheric data, indicates that planetary and cyclonic waves are involved. These waves modify the temperature and wind fields and will be effective in changing the fractional concentration of an ionizable constituent at a given height, if the constituent has a gradient in the vertical or meridional directions.

Gregory and Manson (1970) presented actual examples in support of the hypothesis that planetary wave propagation is responsible for the perturbations of mesospheric electron densities during winter in middle latitudes. Dickinson (1968, 1969) showed that planetary wave propagation through weak eastward flow of the equinoxes is limited, and that maximum wave penetration will occur when wind velocity is below the critical value which reduces the Doppler-shifted wave frequency to zero. Since winds during winter months at middle latitudes are often less than this critical value, large transmission coefficients are possible during that season.

Kellogg (1968) suggested that gravity waves transmitted upwards to the lower thermosphere would "stir" the ionizable constituents of the E region downward into the D region. Kellogg noted that transmission of these gravity waves vertically upward through the stratosphere and mesosphere depends on the wind and temperature fields. During a temporary breakdown of the normal cyclonic vortex of the stratosphere over the winter pole, the "filtering" action of the stratospheric wind field may be eliminated over certain regions because the strong westerly wind prevents planetary-wave disturbances from propagating upwards in winter.

Christie (1970) discussed many common features in the thermal, flow and radiative temperature change distributions in the vicinity of the tropopause (in mid-latitudes) and the winter mesopause (in low latitudes), and he inferred that transport processes may be similar in the two regions. Also, Christie (1970) discussed the transport of nitric oxide in large-scale wave systems, and noted the similarity in the

properties of nitric oxide (in the upper mesosphere). Furthermore, Christie (1970) interpreted the day-to-day variations
in D-region electron densities by analogy with the familiar
tracer transports near and through the tropopause.

Zimmerman and Narcisi (1970) used the temperature and
wind profiles (Sechrist, et al., 1969) measured on normal
and anomalous winter days at Wallops Island to calculate the
Richardson number throughout the D region. The Richardson
criterion is a measure of the presence or absence of turbulence,
and thus an index of turbulent transport. Their results indicated that, on the anomalous day, the mesosphere was fairly
stable ($R_i > 1$) from 60 km up to about 78 km, but between
78 and 90 km (where the electron density was most enhanced)
the mesosphere was turbulent as defined by the condition
$R_i < 1$. However, on the normal winter day the condition for
the existence of turbulence extended over the entire altitude
range from 68 to 86 km. Zimmerman and Narcisi (1970) suggested that there was a *concentration buildup* on the anomalous
day between 78 and 90 km due to rapid transport through the
turbulent region. On the normal day, the whole D region was
very turbulent or in a mixed state from ∿ 68 km upwards, implying that minor species would be well mixed over a larger
altitude range and not concentrated in a narrow interval.

11.5 Recommendations for Future Studies

Ionospheric and airglow observations have indicated
that dynamical processes can cause composition changes at
D-region heights, especially in winter. Two distinct processes
are thought to operate. One is turbulent mixing, the effects
of which can be described by an eddy-diffusion coefficient.
The mechanism of generation of turbulence at these levels is
still a question, but strong vertical shear motions on the
scale of internal gravity waves (up to several hundred kilometers horizontally) are thought to be responsible. The
other process is vertical and horizontal advection by motions
on the planetary-wave scale (several thousand kilometers
horizontally). Which of these processes is the principal
one causing transport of minor constituents in winter is not
known. Information relevant to this problem could be provided by studies of the morphology of the anomalous condition
shown by ionospheric and airglow measurements, by particular
rocket-borne observations in the D region, and by theoretical
studies incorporating realistic transport models (Thomas et.
al., 1969).

a) In order to understand the processes giving rise to
the anomalous behavior of the D region in winter, it is

essential to determine whether the changes in electron
density occur preferentially over certain ranges of height
and, if not, whether the changes at different heights are
related.

b) Since the anomalous behavior of the D region in
winter probably arises from changes in distribution of minor
constituents, such as NO, O ,and H_2O, rocket-borne measurements
of the concentrations of these constituents and of their
vertical transport should be made on anomalous and normal
winter days.

c) The small-scale variations of wind with height
shown by vapor-trail observations should be compared for
anomalous and normal winter days; it would be expected that
this stratification would be modified by a large variation
in vertical transport or eddy diffusion.

d) A primary need in attempting to understand the
effects of dynamical processes in the D region is to distin-
guish, if possible, between changes in composition caused
by transport alone from those caused by changes of chemical
rate constants arising out of changes in temperature.

e) Theoretical studies concerned with the effects of
movements on neutral composition should include vertical
mean motion and horizontal transport.

Little information is available concerning the morphology of the D-region winter day condition. MF and HF absorption measurements could serve as an inexpensive method for
studying the distribution of the winter anomaly aspect of the
condition. There is a need for both regional studies, covering areas of radii up to about 2000 km, and for simultaneous
observations at widely different longitudes, say in North
America and Europe. The establishment of a network of observing sites in North America, together with the stratospheric data provided by the Meteorological Rocket Network and
higher level wind data provided by meteor radar studies,
could provide a good basis for a study of ionosphere-strato-
sphere relationships. The winter anomaly may be due to vertical transport of a minor constituent by eddy diffusion, or
it may be due to vertical or horizontal transport of a minor
constituent by a motion field of planetary wave scale. It
must be stressed that a knowledge of the spatial scale of
the winter anomaly could be a vital step toward the resolution of this question (Thomas et al., 1969).

Since transport processes seem to play a dominant role

in both D region and airglow behavior in winter, it would be useful to have suitable ionospheric and airglow measurements at a number of locations.

The effects of transport processes in the D region seem to be most clearly demonstrated in the anomalous behavior of the region during certain winter days. The following recommendations are, therefore, chiefly directed towards studies of this anomalous behavior (Thomas et al., 1969).

1) The morphology of one or more aspects of the anomalous winter condition should be studied.

2) Studies of the changes in electron density from the normal winter day values should be made simultaneously over the whole height range from lower to upper D region.

3) Rocket-borne measurements of minor constituent concentrations and of positive- and negative-ion composition should be made on anomalous and normal winter days.

4) Attempts should be made to extend the measurements of winds, temperature, and pressure at certain locations in winter to heights above those attained by balloons and meteorological rockets.

5) Ground-based radio and meteor radar studies of winds in the D region should be extended, and attempts should be made to monitor the intensity of D-region turbulence.

6) Attempts should be made to include realistic transport models in theoretical studies concerned with the effects of movements on neutral composition.

11.6   References

Adams, N. G., D. K. Bohme, D. B. Dunkin, F. C. Fehsenfeld and E. E. Ferguson, 1970. "Flowing afterglow measurements of weakly bound cluster ions." Journal of Chemical Physics, Vol. 52, 3133.

Anderson, J. G., 1970. "Rocket-borne ultraviolet spectrometer measurement of OH resonance fluorescence with a diffusive transport model for mesospheric photochemistry." Ph. D. thesis, University of Colorado, Dept. of Astro-Geophysics.

Appleton, Sir Edward, and W. R. Piggott, 1954. "Ionospheric absorption measurements during a sunspot cycle." Journal of Atmospheric and Terrestrial Physics, Vol. 5, 141-172.

Barth, C. A., 1966a. "Nitric oxide in the upper atmosphere." Annales de Geophysique, Vol. 22, 198-207.

Barth, C. A., 1966b. "Rocket measurement of nitric oxide in the upper atmosphere." Planetary and Space Science, Vol. 14, 623-630.

Belrose, J. S., and L. Thomas, 1968. "Ionization changes in the middle latitude D-region associated with geomagnetic storms." Journal of Atmospheric and Terrestrial Physics, Vol. 30, 1397-1413.

Beynon, W. J. G., and K. Davies, 1955. "A study of vertical incidence ionospheric absorption at 2 MHz." The Physics of the Ionosphere, Report of 1954 Cambridge Conference, The Physical Society, 40-52.

Biondi, M. A., 1970. "Recombination coefficients of hydrated positive ions at 400°K." Informal presentation at AGU Meeting in Washington, D. C., April.

Bossolasco, M., and A. Elena, 1963. "Absorption de la couche D et temperature de la mesosphere." C. R. Acad. Sci. Paris, Vol. 256, 4491-4493.

Bourdeau, R. E., A. C. Aikin and J. L. Donley, 1966. "Lower ionosphere at solar minimum." Journal of Geophysical Research, Vol. 71, 727-740.

Bourne, I. A., and L. W. Hewitt, 1968. "The dependence of ionospheric absorption of MF radio waves at mid-latitudes on planetary magnetic activity." Journal of Atmospheric and Terrestrial Physics, Vol. 30, 1381-1395.

Bowhill, S. A., 1969a. "Interactions between the stratosphere and the ionosphere." Annals of the IQSY, Vol. 5, 83-95.

Bowhill, S. A., 1969b. "Ion chemistry of the D- and E-regions --A survey for working group 11 of the Inter-Union Commission on Solar Terrestrial Physics." Journal of Atmospheric and Terrestrial Physics, Vol. 31, 731-741.

Christie, A. D., 1970. "D-region winter anomaly and transport near the mesopause." Journal of Atmospheric and Terrestrial Physics, Vol. 32, 35-56.

CIRA 1965, Cospar International Reference Atmosphere, North-Holland Publishing Co., Amsterdam.

Dickinson, R. E., 1968. "Planetary Rossby waves propagating vertically through weak westerly wind wave guides." Journal of Atmospheric Science, Vol. 25, 984-1002.

Dickinson, R. E., 1969. "Vertical propagation of planetary Rossby waves through an atmosphere with Newtonian cooling." Journal of Geophysical Research, Vol. 74, 929-938.

Dieminger, W., 1952. "On the causes of excessive absorption in the ionosphere on winter days." Journal of Atmospheric and Terrestrial Physics, Vol. 2, 340-349.

Dieminger, W., 1969. "The winter anomaly of ionospheric absorption and meteorological phenomena." Presented at the XVI General Assembly of the International Union of Radio Science, August 18-28, Ottawa, Canada.

Dieminger, W., G. Rose, H. Schwentek and H. U. Widdel, 1968. "The morphology of winter anomaly of absorption." Space Research VII, North-Holland Publishing Co., Amsterdam, 228-238.

Evans, W. F. J., D. M. Hunten, E. J. Llewellyn and A. Vallance-Jones, 1968. "Altitude profile of the infrared atmospheric system of oxygen in the dayglow." Journal of Geophysical Research, Vol. 73, 2885-2896.

Evans, W. F. J., E. J. Llewellyn and A. Vallance-Jones, 1969. "Balloon observations of the temporal variation of the infrared atmospheric oxygen bands in the airglow." Planetary and Space Science, Vol. 17, 933-947.

Evans, W. F. J. and E. J. Llewellyn, 1970. "Molecular oxygen emissions in the airglow." Annales of Geophysique, Vol. 26, 167-178.

Fehsenfeld, F. C. and E. E. Ferguson, 1969. "Origin of water cluster ions in the D region." Journal of Geophysical Research, Vol. 74, 2217-2222.

Ferguson, E. E., 1970a. "Recent results on ion chemistry of the D region of the ionosphere." Annales of Geophysique, Vol. 26, 589-594.

Ferguson, E. E., 1970b. "Laboratory measurements of D-region ion-molecule reactions." Presented at ESRIN-ESLAB Symposium, Frascati, Italy, July 6-10.

Ferguson, E. E. and F. C. Fehsenfeld, 1969. "Water vapor ion cluster concentrations in the D region." Journal of Geophysical Research, Vol. 74, 5743-5751.

Ferguson, E. E. and F. C. Fehsenfeld, 1970. "Recent results on ionospheric positive ion chemistry." Presented at DASA Symposium on Physics and Chemistry of the Upper Atmosphere, June 24-26.

Geisler, J. E. and R. E. Dickinson, 1968. "Vertical motions and nitric oxide in the upper mesosphere." Journal of Atmospheric and Terrestrial Physics, Vol. 30, 1505-1521.

Goldberg, R. A. and A. C. Aikin, 1970. "Equatorial D-region positive ion composition measurements." Presented at Fall AGU Meeting in San Francisco, California, December 7-11.

Gregory, J. B., 1956. "Ionospheric reflections from heights below the E region." Aust. Journal of Physics, Vol. 9, 324-342.

Gregory, J. B., 1965. "The influence of atmospheric circulation on mesospheric electron densities in winter." Journal of Atmospheric Science, Vol. 22, 18-23.

Gregory, J. B. and A. H. Manson, 1969. "Seasonal variations of electron densities below 100 km at midlatitudes--II." "Electron densities and atmospheric circulation." Journal of Atmospheric and Terrestrial Physics, Vol. 31, 703-729.

Gregory, J. B. and A. H. Manson, 1970. "Seasonal variations of electron densities below 100 km at mid-latitudes--III. Stratospheric-ionospheric coupling." Journal of Atmospheric and Terrestrial Physics, Vol. 32, 837-852.

Hesstvedt, E., 1968. "On the effect of vertical eddy transport on atmospheric composition in the mesosphere and lower thermosphere." Geofysiske Publikasjoner Geophysica Norvegica Vol. XXVII, 1-35.

Hesstvedt, E., 1970. "A meridional model of the oxygen-hydrogen atmosphere." Institute of Geophysics, University of Oslo, Oslo, Norway, August.

Hesstvedt, E. and U. B. Jansson, 1969. "On the effect of vertical eddy transport on the distribution of neutral nitrogen components in the D-region." Aeronomy Report No. 32, Aeronomy Laboratory, University of Illinois, Urbana, Illinois.

Huffman, R. E., D. E. Paulsen, J. C. Larrabee and R. B. Cairns, 1971. "Decrease in D-region $O_2(^1\Delta)$ photoionization rates resulting from CO_2 absorption." Journal of Geophysical Research, Vol. 76, 1028-1038.

Hunt, B. G., 1966. "Photochemistry of ozone in a moist atmosphere." Journal of Geophysical Research, Vol. 71, 1385-1398.

Hunten, D. M. and M. B. McElroy, 1968. "Metastable $O_2(^1\Delta)$ as a major source of ions in the D region." Journal of Geophysical Research, Vol. 73, 2421-2428.

Jones, L. M., J. W. Peterson, E. J. Schaeffer and H. F. Schulte, 1959. "Upper air density and temperature: some variations and an abrupt warming in the mesosphere." Journal of Geophysical Research, Vol. 64, 2331-2340.

Kellogg, W. W., 1961. "Chemical heating above the polar mesopause in winter." Journal of Meteorology, Vol. 18, 373-381.

Kellogg, W. W., 1968. "Implications of the stratospheric gravity wave filter to ionization in the D region." Presented at COSPAR (Tokyo).

Lauter, E. A., K. Sprenger and G. Entzian, 1969. "The lower ionosphere in winter." Stratospheric Circulation, editor, W. L. Webb, Vol. 22, 401-438, Academic Press Inc., New York.

Lauter, E. A. and R. Knuth, 1967. "Precipitation of high energy particles into the upper atmosphere at medium latitudes after magnetic storms." Journal of Atmospheric and Terrestrial Physics, Vol. 29, 411-418.

Lauter, E. A. and P. Nitzsche, 1967. "Seasonal variations of ionospheric absorption deduced from A3-measurements in the frequency range 100-2000 kc/s." Journal of Atmospheric and Terrestrial Physics, Vol. 29, 533-544.

Lauter, E. A. and B. Schaning, 1970. "On the low-latitude boundary of the winter anomaly of ionospheric absorption." Journal of Atmospheric and Terrestrial Physics, Vol. 32, 1619-1624.

Maehlum, B., 1967. "On the 'winter anomaly' in the midlatitude D region." Journal of Geophysical Research, Vol. 72, 2287-2299.

Manson, A. H., 1968. "Coupling effects between the ionosphere and stratosphere in Canada (45°N, 75°W) 1962-66." Journal of Atmospheric and Terrestrial Physics, Vol. 30, 627-632.

Manson, A. H. and M. W. J. Merry, 1970. "Particle influx and the 'winter anomaly' in the mid-latitude (L=2.5-3.5) lower ionosphere." Journal of Atmospheric and Terrestrial Physics, Vol. 32, 1169-1182.

Mawdsley, J., 1961. "Air density variations in the mesosphere
and the winter anomaly in ionospheric absorption." Journal of
Geophysical Research, Vol. 66, 1298-1299.

Mechtly, E. A. and J. S. Shirke, 1968. "Rocket electron concentration measurements on winter days of normal and anomalous
absorption." Journal of Geophysical Research, Vol. 73, 6243-
6247.

Mechtly, E. A. and L. G. Smith, 1968. "Seasonal variation of
the lower ionosphere at Wallops Island during the IQSY."
Journal of Atmospheric and Terrestrial Physics, Vol. 30,
1555-1561.

Mechtly, E. A., K. Seino and L. G. Smith, 1969. "Lower ionosphere electron densities measured during the solar eclipse of
November 12, 1966." Radio Science, Vol. 4, 371-375.

Megill, L. R., J. C. Haslett, H. I. Schiff and G. W. Adams,
1970. "Observations of $O_2(^1\Delta)$ in the atmosphere and allowable
values of the eddy diffusion coefficient." Journal of Geophysical Research, Vol. 75, 6398-6401.

Meira, L. G., Jr., 1971. "Rocket measurements of upper atmospheric nitric oxide and their consequences to the lower ionosphere." Journal of Geophysical Research, Vol. 76, 202-
212.

Narcisi, R. S., 1970a. "Shock wave and electric field effects
in D region water cluster ion measurements," Presented at the
Spring 1970 AGU Meeting, Washington, D. C.

Narcisi, R. S., 1970b. "Composition studies of the lower ionosphere." Presented at the International School of Atmospheric
Physics in Erice, Sicily, June 15-29.

Narcisi, R. S. and A. D. Bailey, 1965. "Mass spectrometric
measurements of positive ions at altitudes from 64 to 112
kilometers." Journal of Geophysical Research, Vol. 70, 3687-
3700.

Narcisi, R. S., C. R. Philbrick, C. Sherman, D. M. Thomas, A.
D. Bailey, L. E. Della Lucca, R. A. Wlodyka, G. Federico and D.
Baker, 1970. "D-region composition during disturbed conditions."
Presented at ESRIN-ESLAB Symposium, Frascati, Italy, June 6-10.

Nicolet, M., 1945. "L'interpretation physique de l'ionosphere,"
Mémoires de l'Institut de Royal Météorologie, Belgium., Vol.
19, 106.

Nicolet, M., 1955. "The aeronomic problem of nitrogen oxides." Journal of Atmospheric and Terrestrial Physics, Vol. 7, 152-169.

Nicolet, M., 1962. "Discussion on electron densities in the normal D-layer." Electron Density Profiles in the Ionosphere and Exosphere, editor, B. Maehlum, Pergamon Press, New York, 32.

Nicolet, M., 1965. "Nitrogen oxides in the chemosphere." Journal of Geophysical Research, Vol. 70, 679-689.

Nordberg, W., L. Katchen, J. Theon and W. S. Smith, 1965. "Rocket observations of the structure of the mesosphere." Journal of Atmospheric Science, Vol. 22, 611-622.

Norton, R. B. and C. A. Barth, 1970. "Theory of nitric oxide in the earth's atmosphere." Journal of Geophysical Research, Vol. 75, 3903-3909.

Pearce, J. B., 1969. "Rocket measurement of nitric oxide between 60 and 96 kms." Journal of Geophysical Research, Vol. 74, 853-861.

Potemra, T. A. and A. J. Zmuda, 1970. "Precipitating energetic electrons as an ionization source in the midlatitude nighttime D region." Journal of Geophysical Research, Vol. 75, 7161-7167.

Reid, G. C., 1970a. "Production and loss of electrons in the quiet daytime D region of the ionosphere." Journal of Geophysical Research, Vol. 75, 2551-2562.

Reid, G. C., 1970b. "The roles of water vapor and nitric oxide in determining electron densities in the D region." Presented at ESRIN-ESLAB Symposium, Frascati, Italy.

Sechrist, C. F., Jr., 1967. " A theory of the winter absorption anomaly at middle latitudes." Journal of Atmospheric and Terrestrial Physics, Vol. 29, 113-136.

Sechrist, C. F., Jr., E. A. Mechtly, J. S. Shirke and J. S. Theon, 1969. "Coordinated rocket measurements on the D-region winter anomaly 1. Experimental results." Journal of Atmospheric and Terrestrial Physics, Vol. 31, 145-153.

Sechrist, C. F., Jr., 1970. "Interpretation of D-region electron densities." Radio Science, Vol. 5, 663-671.

Shapley, A. H. and W. J. G. Beynon, 1965. "Winter anomaly in ionospheric absorption and stratospheric warmings." Nature Vol. 206, 1242-1243.

Shimazaki, T. and A. R. Laird, 1970. "A model calculation of the diurnal variation in minor neutral constituents in the mesosphere and lower thermosphere including transport effects." Journal of Geophysical Research, Vol. 75, 3221-3235.

Strobel, D. F., D. M. Hunten and M. B. McElroy, 1970. "Production and diffusion of nitric oxide," Journal of Geophysical Research, Vol. 75, 4307-4321.

Theon, J. S., W. Nordberg, L. B. Katchen and J. J. Horvath, 1967. "Some observations on the thermal behavior of the mesosphere." Journal of Atmospheric Sciences, Vol. 24, 428-438.

Thomas, L., 1962. "The winter anomaly in ionospheric absorption." Journal of Atmospheric and Terrestrial Physics, Vol. 23, 301-317.

Thomas, L., 1968. "The electron density distribution in the D and E regions during days of anomalous radio wave absorption in winter." Journal of Atmospheric and Terrestrial Physics, Vol. 30, 1211-1217.

Thomas, L., J. F. Bedinger, J. E. Geisler, C. O. Hines, T. Shimazaki, J. S. Theon and S. P. Zimmerman, 1969. "Transport processes in the D region,"STP Notes No. 5, Inter-Union Commission on Solar-Terrestrial Physics, 43-46.

Tulinov, V. F., 1967. "On the role of corpuscular radiation in the formation of the lower ionosphere (below 100 km)." Space Research VII, North-Holland Publishing Co., Amsterdam.

Tulinov, V. F., L. V. Shibaeva and S. G. Jakovlev, 1969. "The ionization of the lower ionosphere under the influence of corpuscular radiation." Space Research IX, North-Holland Publishing Co., Amsterdam, 231-236.

Vallance-Jones, A. and R. L. Gattinger, 1963. "The seasonal variation and excitation mechanism of the 1.58 $^1\Delta_g$-$^3\Sigma^-$ twilight airglow band." Planetary Space Science, Vol. 11, 961-974.

Webber, W. R., 1962. "The production of free electrons in the ionospheric D layer by solar and galactic cosmic rays and the resultant absorption of radio waves." Journal of Geophysical Research, Vol. 67, 5091-5106.

Weeks, L. H., 1967. "Lyman-alpha emission from the sun near solar minimum." The Astrophysical Journal, Vol. 147, 1203-1205.

Wood, H. C., W. F. J. Evans, E. J. Llewellyn and A. Vallance-Jones, 1970. "Summer daytime height profiles of $O_2(^1\Delta)$ concentration at Fort Churchill." Canadian Journal of Physics, Vol. 48, 862-867.

Young, C. and E. S. Epstein, 1962. "Atomic oxygen in the polar winter mesosphere." Journal of Atmospheric Science, Vol. 19, 435-443.

Zimmerman, S. P. and R. S. Narcisi, 1970. "The winter anomaly and related transport phenomena." Journal of Atmospheric and Terrestrial Physics, Vol. 32, 1305-1308.

CHAPTER 12

PHOTOCHEMICAL MODELS IN UPPER ATMOSPHERIC RESEARCH

Eigil Hesstvedt

University of Oslo, Norway

12.1 Introduction

The chemical composition of the upper mesosphere and the lower thermosphere provides a problem of great interest for the study of physical phenomena occurring in this region. It is well known that the distribution of neutral species to a large extent governs the formation of the D-layer. This alone gives us sufficient reason to analyze carefully this part of the upper atmosphere.

Direct measurements are very difficult to perform in this region, and the results are in many cases difficult to interpret. They are also too scattered in space and time to provide us with a satisfactory picture. In order to increase the information from each experiment and also to formulate future problems, photochemical atmosphere models are used. A selected set of chemical reactions is assumed to simulate the photochemistry of the real atmosphere, and on this basis a theoretical composition is computed. Such models have proven very useful in upper atmospheric study, and many questions have been answered by such an approach.

However, in many cases the results from the models do not agree with experiments. The reason for this may be twofold: the chemistry of the model possibly gives an inadequate description of the problem we want to study. In most cases the discrepancy arises from the fact that the chemical composition at a point depends not only on the chemistry, but also on two mechanisms, photochemistry and transport. Only when these two mechanisms are considered together can we expect to obtain a realistic model.

12.2 Computational Methods

Preliminary computations with purely photochemical models show that horizontal gradients in the number densities are relatively small, so that the composition is influenced to only a smaller extent by horizontal transport (except, of course, in high latitudes in winter, where horizontal transport into the dark region is extremely important). And since so little is known about the vertical mean motion, the models have so far been restricted to considering vertical diffusion (molecular and turbulent). The result is that the theoretical distribution of chemical components obtained from such a model will be changed to only a small extent when vertical mean motion and horizontal transport, by mean motion and by eddies, is considered as well.

The vertical flux (F) of a component with number density (n) is given by

$$F = -D\left(\frac{\partial n}{\partial z} + \frac{n}{H} + \frac{n}{T}\frac{\partial T}{\partial z}\right) - K_z\left(\frac{\partial n}{\partial z} + \frac{n}{H_a} + \frac{n}{T}\frac{\partial T}{\partial z}\right) \qquad (12.1)$$

where D is the molecular diffusion coefficient, K_z is the vertical component of the turbulent diffusion coeffient, H is the scale height of the component, H_a is the scale height of the environment, and T is the air temperature. The continuity equation for the component then takes the form

$$\frac{\partial n}{\partial t} = P - Q \cdot n - \frac{\partial F}{\partial z} \qquad (12.2)$$

where P is the production term and Q·n is the destruction term due to photochemical processes.

12.3   The Photochemical Model

Equation 12.2 may be used to make a time dependent computation of the chemical composition in a vertical column, starting from an arbitrary distribution and continuing until a 24 hour period repeats itself. Many models extend so high that diffusive equilibrium may be used as the upper boundary condition. If the lower boundary is taken at 65 km or lower, one may assume photochemical equilibrium as the boundary condition. Such a procedure requires long computer time since the time steps must be short and the convergence is relatively slow. However, for many problems it is possible to simplify the computation considerably and still get an accuracy which is sufficient for that particular study. Examples of problems where such simplifications are recommended are the profiles of atomic oxygen, nitric oxide and water vapor in the upper mesosphere and lower thermosphere.

Water vapor has a long lifetime and very small diurnal variation. Therefore, the flux term will be almost the same day and night. If we take the diurnal means of the P- and

Q-terms, we have a steady state problem which requires very short computer time. We may therefore repeat the computation for a variety of profiles of K_z and also estimate the effect of various values of the vertical mean motion. Similarly, a simplified computation of atomic oxygen gives good approximations down to about 80 km; below this level photochemical equilibirum is a fair assumption. For nitric oxide the simplified model may be used down to 50 km (if we consider the sum of NO and NO_2 rather than NO alone).

Computations of profiles of atomic oxygen and ozone are shown in Fig.12.1. A logarithmic profile is used for K_z, with values of $5 \times 10^6 cm^2/s$ at 105 km and $6 \times 10^6 cm^2/s$ at 80 km. It is seen that the diurnal variation of $O(^3P)$ is small in the thermosphere, while there is a sudden breakdown after sunset and regeneration after sunrise in the mesosphere. The maximum at 90-95 km is due to eddy transport from above. A weaker transport would lower the peak value and also lift the zone of strong increase with height (in the diagram at 80-85 km) by 2-3 km. However, even considerable changes in K_z will not cause drastic changes in the oxygen profile.

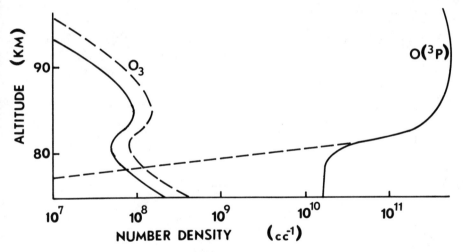

Fig. 12.1. Theoretical profiles of atomic oxygen and ozone in a photochemical model with vertical diffusion (solid curves: daytime values, broken curves: nighttime values).

In the mesosphere and thermosphere hydrogen will mainly be present in the form of H_2O, H_2 or H. Figure 12.2 gives a picture of the broad lines in this distribution. Below 80 km H_2O is, in competition with H_2, the most important hydrogen component. At higher levels H becomes more and more important and dominates the hydrogen budget above 100 km.

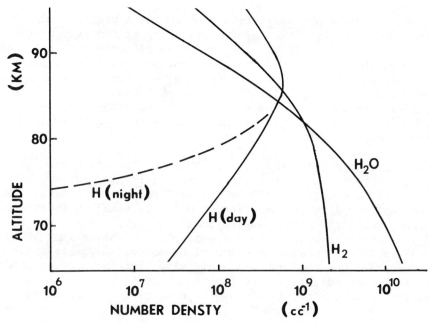

Fig. 12.2.   Theoretical profiles of H_2O, H_2 and H.

Figure 12.3 shows a theoretical profile of NO, computed by Isaksen for 45°, summer.   A neutral atmosphere model would give unrealistic results above 90 km, where $N_2^+ + O \rightarrow NO^+ + N$ is the most important source of NO.

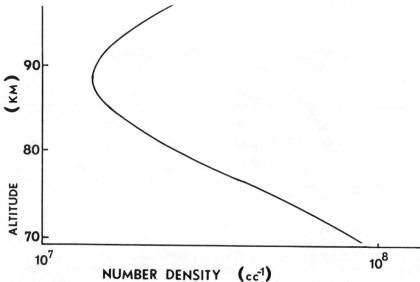

Fig. 12.3.   Vertical profile of NO (45°, summer) computed by Isaksen (unpublished).

CHAPTER 13

AIRGLOW

Rufus E. Bruce

Physics Department,
University of Texas at El Paso

13.1 Introduction

The term *airglow* is used to describe the low intensity, rather steady, radiation emitted by the upper atmosphere. The sudden sky brightenings and other phenomena of the aurora are not included in airglow. Dayglow, twilightglow and nightglow refer to the airglow observed during those specific times. This illumination at ground level is equivalent to that produced by one candle at 100 meters distance (Bates, 1960).

Airglow, as with most radiative processes, is dependent upon the physical conditions in its emission region. Consequently the observed airglow intensities and their fluctuations can be related to the dynamics of the upper atmosphere. This relationship between airglow intensities and upper atmosphere physical processes will be reviewed in this paper.

13.2 Prominent Spectral Features

Much of the earlier research of airglow was devoted to identifying the observed *spectra*. For the most part, the spectra of nightglow and twilightglow contain similar characteristics. However, the observed intensities differ substantially.

The most prominent *nightglow* emission line is the forbidden oxygen transition, $\lambda = 5577.348$ Å. The radiant energy content of this line accounts for 6% to 9% of the total visual nightglow in the atmospheric spectrum. This line was first measured by Babcock in 1923. Later, Slipher (1929) obtained and Paschen (1930) identified the other atomic oxygen forbidden transitions in the airglow: $\lambda = 6300$ Å and $\lambda = 6364$ Å. Slipher (1929, 1933) also found the sodium D lines at $D_1 = 5895.92$ Å and $D_2 = 5889.95$ Å.

The forbidden Herzberg bands of O_2 were identified in the nightglow during the 1940's. Meinel (1948, 1950) identified the O_2 atmospheric system and later the Meinel Band of OH in the visible and near infrared nightglow. The preponderance of all airglow emission takes place in the OH bands, predominately in the near IR region. Rocket measurements during the 1950's confirmed the presence of Lyman α. Hα emission of approximately 1/10 the brightness of OI-6300 was observed by Prokudina (1959).

Twilightglow is the emission observed from below a sunlight layer when the intervening atmospheric layers are in the earth's shadow. The geometry of twilight observations is shown in Fig. 13.1. Brightnesses and thicknesses of emitting layers are obtained from the variation in brightness as shadow height increases. Hunten and Shepherd (1954) used the rapid dropoff of NaD intensity, as shadow height exceeded 85 km, to demonstrate the presence of the 85 km sodium emission layer.

Resonant scatter of solar radiation enhances the emission in several airglow lines: NaD, OI-6300 and several others. Total abundance calculations of sodium based only upon resonant scatter processes agree with that obtained from the terrestrial component of D-lines absorption in the solar spectrum. However, totally unacceptable scale heights of OI would be required to explain the twilightglow enhancement of OI-6300 by resonant scattering processes.

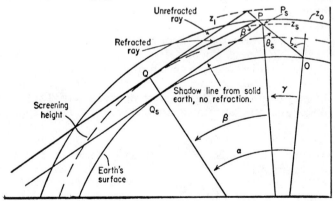

Fig. 13.1. Two-dimensional *geometry* of twilightglow. The refracted incident ray that passes above the screening height intersects the line of sight (zenith angle ζ) at P. The atmosphere below z_S is in the earth's shadow. The angles β and γ are not in the same plane. (From Chamberlain, 1961.)

The intensity of the OI-6300 twilightglow enhancement, given in
Fig. 13.2, is 10 to 20 times its average nightglow value. Near
mean values of intensity are not reached for several hours.
This slow reduction and a corresponding predawn inhancement indicate that other mechanisms are in operation.

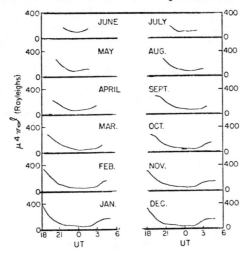

Fig. 13.2. OI-6300 Nightglow. Mean monthly curves for the
daily variation show the slow post–twilight decay and the predawn enhancement. From Chamberlain (1961), based on observations between May 1957 and April 1960 due to D. Barbier.

     Twilight observations of OH indicate an emitting layer in
the region of 85 km for which OH band intensity is related to
emission height. The predicted wavelength and band intensities
of OH are given by Chamberlain (1961) and Roach (1968).

     In addition to the preceding, the H and K lines of calcium
II, the 7699 Å line of potassium, and the resonance lines of
lithium I have also been observed in the twilight glow. The
height of the lithium and potassium emitting layer is similar
to that of sodium. Calcium emission layers are at 100 and 250
km. Sheflov (1961) observed the 10,830 Å He auroral line at
extreme heights of between 500 and 1500 km in the twilight
glow. The 5199 Å nitrogen line is also found in the twilightglow.

     The large scattered solar radiation make ground obser-
vation of the dayglow extremely uncertain. Recent rocket and
satellite measurements have established the presence of
Hydrogen Lyman α and Hα in the dayglow (Fastie, 1968).

13.3 Airglow Emission Brightnesses

The accepted unit of absolute airglow *intensity* is the raleigh (R) which is defined as follows: 1R = apparent emission rate of 10^6 photons cm^{-2} (column) sec^{-1}. The intensity is the measured brightness (B) multiplied by 4π (B may be corrected for lower atmospheric scattering and extinction). From the energy standpoint 1R of Lyman α is equivalent to about 10R at 10,000 Å. The intensities of the principal nightglow spectral features are listed in Table 13.1. The information is based upon data published by Krassovsky, Shefov and Yarin (1962) and given by Roach (1968). A summary of twilightglow data is given in Table 13.2 which is based upon data given by Kvifte (1969).

13.4 OH Contamination

The predominate emitter (predicted) in the airglow is OH. Total nightglow integrated intensity of OH (in R) is 1000 times greater than all other emissions combined; its quantal radiation is equivalent to the quantal intensity in the visable mid-twilight sky. (Roach (1968) points out that OH emissions in the same spectral region as OI-5577, OI-6300 and NaI-5890 have caused intensity errors. As an illustration, the uncorrected error in the OI-6300 intensity can be as high as 80%, depending upon the amount of OH contamination. Photometric observations require that considerable care be taken to prevent OH contamination.

13.5 Airglow *Excitation Processes*

Of the possible energy sources, four that affect airglow are:

  a. High energy particle collisions.
  b. Solar radiation.
  c. Chemical reactions.
  d. Thermal collisional excitations.

Solar radiation and chemical reactions are the important airglow energy sources. High energy collisions are related to auroral displays, and thermal collisions are important primarily in quelching and thermalizing processes.

Regardless of the energy source all processes must satisfy four major requirements for prominent atmospheric emission (Swider, 1969). They are:

1. The process must be exothermic.
2. There must be a substantial exciting particle (or photon) flux and target atomic or molecular concentration.
3. The process must be reasonably efficient.
4. Quenching lifetimes should not be substantially less than radiative lifetimes.

Table 13.1   Nightglow intensities[a]

Species	Wavelength	Zenith Intensities (absolute) $(10^{-4}$ erg cm^{-2} sec$^{-1})$
OH	0.38 μ to 4.5μ (Meinel)	3.6×10^4 (5,000,000 R)
O_2	8645 Å (atmospheric)	11 (500 R)
	3000-4000 Å (Herzberg)	88 (1500 R)
OI	6300, 6364 Å	6.2 (200 R)
	5577 Å	8.9 (250 R)
HI (H)	6563 Å	0.45 (15 R)
(H)	4861 Å	1.2 (3 R)
NaI	5890, 5896 Å[b] Summer[b]	1.0 (30 R)
	5890, 5896 Å[b] Winter[b]	6.8 (200 R)
Continum Nightglow	4000-7000 Å	50 (1500 R) Mean 0.5 R/Å
Astronomican	4000-7000 Å	150 (4500 R) Mean 1.5 R/Å

[a] After Roach (1968).
[b] D_1/D_2, intensity ratio is 2.1.

Many of the airglow emissions from lower levels of oxygen and nitrogen appear to violate the third requirement since the radiation originates from metastable levels. The lower levels and transition data for atomic oxygen and nitrogen are shown in Fig. 13.3; the corresponding data for the molecular species are shown in Fig. 13.4. The argument in favor of observing forbidden transitions is that quenching rates are equally slow. Upper limits for quenching rates can be obtained from transition probabilities (Swider 1969).

Table 13.2 Twilightglow summary[a]

#	Atom or Molecule	Line or Band	$\lambda(\text{Å})$	Brightness Rayleighs (R)	Peak Concen. Height (km)	Chamberlain[b] Maximum Brightnesses
1.	Li I	-	6708	10-200	85-90	200 R
2.	Na I	$D_1 D_2$	5890/96	$1\text{-}5 \times 10^3$	90	1000(summer)-5000(winter)
3.	K I	-	7699	30	88	-
4.	Ca II	K	3933	-	100-	150 R
5.	-	H	3967	$10\text{-}10^2$	250	-
6.	He I	-	10830	$0.1{=}1{\times}10^2$	500+	-
7.	-	-	3889	1	-	-
8.	N_2+	1.Neg0-0	3914	50-500	300	-
9.	"	" 0-1	4278	-	-	-
10.	"	" 1-2	4236	-	-	-
11.	"	" 2-3	4199	-	-	-
12.	N_2+	Meinel 1-0	9200	-	-	-
13.	N I	$(NI)_{21}$	5200	20	300?	-
14.	O I	$(OI)_{21}$	6300	100-200	300-350	1000 R
15.	O I	$(OI)_{32}$	5577	50-500	200	-
16.	O_2	Atmos.0-1	8645	-	50-100	-
17.	O_2	Infrared Atmos.0-1	15803	$200{\times}10^3$	50-100	20,000 R

[a] From B. Kvifte (1969).
[b] From Chamberlain (1960).

Basically the excitation processes for nightglow and twilightglow are different. Nightglow is caused primarily by chemical processes while twilightglow results from interactions with solar radiation. Twilight processes will be discussed first, followed by nightglow processes, and then by a section on temperature determinations.

Twilight emission spectra outlined in Table 13.2 result from interactions with *solar radiation*, usually in one of the following type reactions:

$$A + h\nu_0 \rightarrow A + h\nu_0 \quad \text{- resonance radiation} \tag{13.1}$$
$$\rightarrow A^* + h\nu_1 \quad \text{- flourescent radiation} \tag{13.2}$$

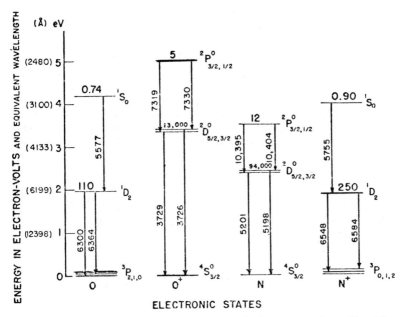

Fig. 13.3 Atomic states and transitions wavelength (from Swider, 1969).

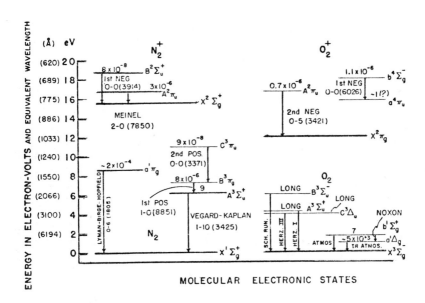

Fig. 13.4 Molecular states and transition wavelengths (from Swider, 1969).

$$AB + h\nu_0 \rightarrow A + B + h\nu_2 - \text{photodissociation} \tag{13.3}$$

$$A + h\nu_0 \rightarrow A^+ + e + h\nu_3 - \text{photoionization} \tag{13.4}$$

$$A + e \rightarrow A^* + e \rightarrow A + e + h\nu_4 - \text{collisional excitation} \atop \text{(using photoelectron)} \tag{13.5}$$

Process (13.1) is the mechanism for most twilightglow excitation. The observed radiation of each of the following is consistent with resonance or flourescent processes LiI-6708, NaI-5890/96, KI-7699, CaII-3968, HeI-3889, and the O_2 atmospheric 0-1 band.

The red and green atomic oxygen lines are excited during twilight through photodissociation of O_2 (Deehr, 1968). An excitation mechanisms for the O_2 infrared atmospheric 0-1 band also appears to be process (13.3):

$$O_3 + h\nu_0 \rightarrow O + O^*_2 \rightarrow O + O_2 + h\nu \tag{13.6}$$

Broadfoot, (1967b, 1968), Hunten (1964, 1967) and Broadfoot and Hunten (1966) account for the excitation and necessary ionization associated with the first negative band system of N_2^+ through the following:

$$N_2 + h\nu \rightarrow N_2^+ + e \text{ , and} \tag{13.7}$$

$$N_2^+ + h\nu_0 \rightarrow N_2^+ + h\nu_0 \tag{13.8}$$

photoionization followed by resonant scatter.

The two twilightglow He lines are also two step processes. They are respectively transitions from the 2^3P and 3^3P levels to the metastable 2^3S level. Both observed lines are thought to be resonant excitation from the 2^3S level following collisional excitation of the metastable level from the ground state by atmospheric photoelectrons.

Predawn enhancements are also explained through process (13.5). Sunlight at the conjugate magnetic point in one hemisphere produces photoelectrons which migrate to the opposite hemisphere along field lines and enhance radiation through collisional excitation. In the case of OI-6300 predawn enhancement, Cole (1965) proposed that conjugate point electrons elevated temperatures until $e + O \rightarrow O(^1D) + e$ becomes important. The absence of OI-5577 enhancement indicates only low energy electrons are involved.

Quenching processes, probably involving N_2, are important in the OI-6300 (red line). Collisional quenching of the long lifetime state explains why OI-5577 is observed in lower altitudes but not OI-6300 (see Fig. 13.3). Quenching does not appear to be important in OI-5577 over about 100 km altitude.

Photochemical processes are the main source of nightglow. Only H and He emissions do not involve photochemical processes; it appears that they involve transport and resonance scatter of solar radiation from sunlight regions.

Because of the many different possible chemical reactions involving the same or similar species, no single process can be identified as the only possible excitation process for a given emission. The atmospheric chemical reactions involve both electrically neutral and ionic species. In general there are always competing processes that excite a given nightglow emission. Such is the case with sodium. Chapman (1939) proposed $Na + O \rightarrow Na(^2P) + O_2$ as the excitation process for the sodium D lines. Bates (1960) found a competing process necessary: $NaH + O \rightarrow Na(^2P) + OH$. Competing processes appear probable for OH Meinel band excitation (Packer, 1961): $O_2^* + H \rightarrow OH^* + O$ (Krassovsky and Sefov, 1965); and $O_3 + H \rightarrow OH^* + O_2$ (Bates and Niolet, 1950).

Cetain processes can lead to a variety of product states. For instance the three body reaction, $O + O + M \rightarrow O_2^* + M$, leads to any number of excited O_2 states, O_2^*. Dissociative recombination of O_2^+, $O_2^+ + e \rightarrow O + O + Energy$ leads to five different atomic oxygen excited state combinations. Similarly NO^+ dissociative recombination leads to 3 different channels of the form $NO^+ + e \rightarrow N + O + Energy$.

The Chapman process for the OI-5577 line (Chapman, 1931, 1937) is important in explaining the 100 km oxygen layer:

$$O + O + O \rightarrow O_2 + O(^1S) + 0.95 \text{ ev} \qquad (13.9)$$

A 10% recombination of atomic oxygen during the night in the above process is sufficient to produce a 300 kR nightglow.

Emission of the OI-5577 line at 100 km without the OI-6300 line requires quenching of $O(^1D)$. Swider (1969) has proposed $O_2 + O(^1D) \rightarrow O_2(b'\Sigma_g^+) + O$ as a mechanism to quench the OI-6300

radiation and at the same time account for the O_2 atmospheric
bands in the nightglow. (Also see Young and Black, 1966.)

The absence of OI-5577 in the M arcs while it is present
in the ionosphere F-region layer implies that entirely different processes are operative in the two regions. The requirement is for a low energy excitation process such as one of the
following: a) exothermic photochemical reaction; b) electrical
discharge; or c) a thermal excitation (Megill and Carlton,
1964, Cole, 1965).

One of the basic differences between airglow and aurora
is the excitation mechanism. Aurora results from collision
excitations caused by an incident high energy particle flux.
More detailed reviews of excitation processes have been made
by Chamberlain (1961) and Swider (1969) which provided the
basis for this section.

In the upper atmosphere rotational and translational collision rate coefficients are high enough to cause rotational
energy level populations and translational kinetic energies
to be thermally distributed. McPherson and Jones (1960) point
out that the collision frequency for OH molecules at 90 km is
1.9×10^4 sec^{-1} and that the average lifetime of an excited
state is 10^{-2} sec. The average of 200 collisions per excited
molecule assures equilibrium in the rotational levels. Based
upon thermal equilibrium, *temperatures* may be obtained from
the intensity of the rotational lines. The line intensity is
given by

$$I = Ci(J^1) \exp -F(J^1)/kT \qquad (13.10)$$

where J^1 is the rotational quantum number of the upper level,
C is a constant, $i(J^1)$ is the intensity factor and $F(J^1)$ is the
rotational energy of the upper level. Early OH band temperatures obtained at midlatitudes were on the order of 220°K,
which is correct for emission at 85 km. However, in 1953
Chamberlain and Oliver discovered OH temperatures of 300°K
over northern Greenland. Subsequently, OH temperatures above
60° latitude were reported to be as high as 460°K (Prokudina,
1959). Kvifte (1967) observed that the high latitude variations in OH temperature resembled the seasonal variation in
temperature at 100 km. He concluded that the emission layer
for OH at high latitude was near 100 km. Recent rocket
measurements locating the OH emission peak just under 100 km.
Recent rocket measurements locating the OH emission peak just
under 100 km confirm Kvifte's conclusions (Stair and Gauvin,
1967; Baker and Waddoups, 1967, 1968). OH emission temperatures are now routinely obtained.

Kaplan-Meinel band rotational temperatures for O_2 (Temperature between 130°K and 200°K) indicate emission heights at the temperature minimum near 82 km.

In addition to rotational temperatures, observation of doppler line contours can lead to prevailing temperatures. Temperatures deduced from the doppler width of the OI-5577 and OI-6300 lines are in agreement with their observed emission heights.

13.6 Airglow Morphology

In the remaining pages the properties of the observed airglow will be related to atmospheric physical processes. The geographic location and spatial and temporal variations of the airglow will be reviewed.

A schematic representation of the geographic and altitude distribution of some of the nightglow emitting layers is given in Fig. 13.5. Also shown in Fig. 13.5 is the M Arc region, which is related to the auroral activity.

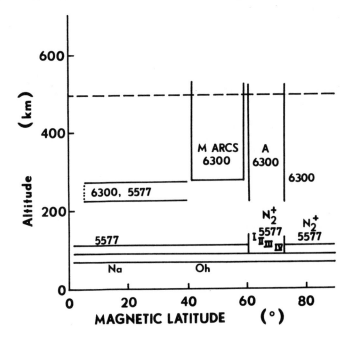

Fig. 13.5 Schematic representation of important nightglow features. Magnetic latitude and altitude of emitting layers is given (from Roach et al., 1967).

The following general observations may be made concerning airglow phenomena:

a) Most emissions show seasonal, annual, diurnal and semiannual variation,

b) They also show latitudinal effects, most of which are believed to be magnetically controlled, and indicate different basic excitation processes in operation.

c) Comparison of the IGY and IQSY indicate that intensities can be related to the sunspot cycle.

d) Some emission shows enhancement during twilight and before dawn.

Most of the airglow data systematically collected during the IGY and IQSY consists of intensities of OI-5577, OI-6300, NaI-5890/96 and OH bands. As a consequence, much more is known about these particular species. A summary is given of the characteristics of the important airglow emitters.

Outside of auroral emissions there are two principal atomic oxygen layers; one is located at about 100 km in which OI-6300 is absent and the other layer is at 240 km in which both OI-5577 and OI-6300 are observed.

In the lower of these two layers a midlatitude maximum in the 5577A intensity is observed to change position over the course of the year. This is evidenced in the so-called north/south ratio data. This ratio of the intensity north of the observation point to that observed south of the station exhibits a distinct semiannual variation. A typical variation is shown in Fig. 13.6. North/south ratios indicate a distinct magnetic dependence (Christophe-Glamme, 1965). Roach, Smith and McKennan (1969) point out that these variations are truly seasonal; similar variations in the southern hemisphere occur with a 6-month time shift, as shown in Fig. 13.7. The two maxima are definitely not symmetrical with respect to season. Tohmatsu and Nagata (1963) have related the change in location of the midlatitude maximum of OI-5577 to large scale meteorological transport of oxygen allotropes.

The upper oxygen layer, centered about 240 km, extends from the equatorial region into midlatitude. The following have been observed about this layer.

a) The intensities show considerable temporal changes.

b) The layer tends to structure itself into large photometrically detected patches (Steiger et al., 1966; Roach and Smith, 1967).

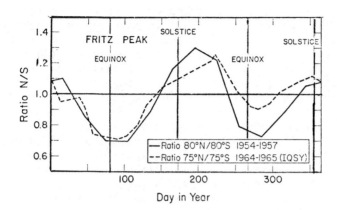

Fig. 13.6   Mean OI-5577 intensity ratios, variation of north/south ratio with day in the year for Fritz Peak during the periods shown (From Roach et al., 1969).

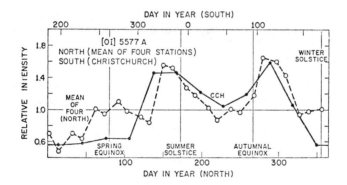

Fig. 13.7   Comparison of the seasonal variation of OI-5577 intensity in the southern hemisphere with the northern hemisphere data (From Roach et al., 1969).

      c)   Optical phenomena are associated with the equatorial ionospheric anomaly.
      d)   The brightness of OI-6300 to OI-5577 is in the ratio of 4:1.

The other prominent oxygen emission area is the M Arc region. The absence of OI-5577 in this region suggests an activation process different than that of the 240 km layer. Their frequency of observation is strongly related to sunspot activity and their intensity is positively correlated to geomagnetic activity. All of these suggest collisional processes more characteristic of aurora rather than airglow.

Silverman (1969) and Brenton and Silverman (1970) have studied the latitudinal dependence of the OI-5577 diurnal variation. Stations north of 45° latitude showed no statistically significant diurnal variation; stations in equatorial regions were characterized by a night variability with a pronounced minimum during the night; and stations between 20°N and 45°N indicate a maximum occurring during the night. These variations have been related to vertical movement of ionization. They correlate with changes in ionosphere virtual heights (h).

All three *alkali metal* emissions (sodium, lithium and potassium) are centered at 85 to 95 km; however, their emission characteristics are quite different. Sodium twilightglow exhibits the following properties:

a)   Above 50°N latitude a pronounced winter maximum is observed which disappears for latitudes less than 35°N. Similar local winter phenomena is observed in the southern hemisphere.

b)   The thickness of the layer at half intensity is about 20 km although thinner layers have been observed.

c)   The diurnal variation of the nightglow intensity of NaI-5890 shows a latitudinal and seasonal dependence (Silverman, 1965; Barbier 1959).

d)   Sodium abundances of 5 to 10 x 10^9 free atoms cm^{-2} column have been deduced from twilight data.

Properties of the natural lithium airglow have been difficult to determine due to the injection of lithium into the atmosphere by nuclear detonation. There is an annual LiI-6708 intensity variation quite different in character from that of sodium (Hunten, 1967). Intensities can be related to meteor showers. The natural atmospheric lithium abundance appears to be about 5 x 10^6 cm^{-2} column and has a distribution similar to that of sodium.

Potassium with abundances of 5 x 10^7 cm^{-2} column exhibits no annual variation in twilightglow   (Kvifte, 1969).

Shefov and Truttse (1969) have summarized the OH emission as follows:

a) High resolution measurements show OH bands with V=2.
b) OH emission shows a sunspot cycle correlation. An emission maximum was noted in 1958 with a small minimum in 1961 and 1962.
c) The annual variation has a winter maximum and a summer minimum.
d) The amplitude of the seasonal variations of rotational temperature increases for bands from higher initial vibrational levels.
e) Periods of 27 and 29.5 (moon's phase) have been detected in the mean diurnal intensities and in rotational temperatures. These variations have been correlated with simultaneous mean atmospheric temperatures at the 50-mb level and with 15-mb and 10-mb iosbaric surfaces.
f) Short period temperature variations with amplitudes of 30 to 40°K are observed.
g) Inphase and synchromous variations of the average rotational temperature have been observed at distances up to 3000 miles (Fedorova, 1967; Sidorov, 1968).
h) OH intensity, along with that of the (0-1) O_2 atmospheric band, increases during the appearance of noctilucent clouds and sharply decreases the following night. On the succeeding night average intensities are again observed. One concludes that OH intensity is affected by atmospheric dust content.
i) OH intensities reach a sharp minimum three or four days after meteor showers: dust content reduces OH intensity.

Hydrogen intensities show annual, seasonal and diurnal variations. The substantial changes from night to night indicate competitive excitation processes. The observed variations in Hα and Lyα are caused in the near earth hydrogen. The maximum yearly mean in Hα of 20R was observed in 1962. The morning side intensity of Hα is greater than the evening side intensity.

The H and K lines of singly ionized calcium are observed only from May to August. Ion distribution which extends from 100 to 250 km does not appear to be homogeneous. Most evidence points to meteorite injection of calcium.

Correlations exist between solar activity and the emission of several atmospheric species; oxygen, sodium, hydroxel and

others. Silverman (1969) suggests that the dynamics of the
90 km region are affected by solar activity. The most obvious
coupling mechanism in this connection is the solar ultraviolet
radiation.

13.7 Airglow as an Ionospheric Probe

Ionospheric phenomena can be related to the airglow through
excitation processes. Since different species radiate from
different layers, the altitude dependence of the phenomena can
also be studied.

Recent papers by Thomas (1969) and Roach (1969) summarize
many of the known relationships between airglow and the ionosphere. A few of these correlations will be reviewed.

The 240 km OI-6300 intensity is related to electron concentrations through its excitation process (dissociative recombination of O_2^+). In explaining the intertropical arcs, OI-
6300 emission layers near 15° North and 15° South, Barbier
(1961) introduced a semi-empirical equation which relates OI-
6300 intensity, Q, to F-layer parameters:

$$Q = Kf^2_o \ (h' + H)e - \frac{h' - 200}{H} + C \qquad (13.11)$$

where K and C are constants, h' is virtual altitude, H is
scale height and $f_o(h' + H)$ is critical frequency of the F
layer at a height H above its base. A somewhat simpler equation,
also in use, related f_o to the height of maximum electron concentration. Large scale variations in midlatitude and tropical regions have been explained using Barbier's equation. Short
scale, fine structure has not been successfully explained.

Roach and Smith (1967) suggest that temporal and spatial
perturbations of 240 km OI-6300 is related to guidance of
ionization down magnetic tubes to the F-layer.

Roach (1969) in his review of the tropical ionosphere
shows that the intensity of OI-6300 is maximum   when h'F is
a minimum. His findings suggest that time coincidences of
minimum virtual height and greatest rate of loss of electrons
in the F-layer is a property of the tropical ionosphere. He
concludes that there must be a downward movement of ionization.
These vertical ionospheric drifts on the order of 10 mps were
theoretically predicted by Maeda (1962) and have been observed
at Jicamarca to be upward in daytime and downward at night.

Abbot (1964) and Dachs (1967, 1969) have shown a correlation between OI-5577 variations and sporadic E. Variations

in intensity of OI-5577 and sporadic E virtual height, h^1, are inversely related: intensity increases with an apparent downward movement. Dachs (1969) states that vertical movements in the upper atmosphere may be studied from 5577A nightglow intensity measurements. Typical curves illustrating these variations are given in Fig. 13.8 (due to Dachs, 1969).

The airglow of species other than oxygen are also of great value as probes of the upper atmosphere. For instance, the Na and OH emissions can be related to solar UV. The airglow of minor species in the D region is extremely important since species with abundances of 1 part in 10^5 can drastically change the chemistry of the region. Doppler shifts of airglow lines, related to mass movements in the order of 100 mps, have been observed (Armstrong, 1969). Radial velocities up to 40 km/sec (doppler measurements) have been deduced from observations of Hβ emission (Reay and Ring, 1969). Hesstvedt (1970, 1969) has studied the effect of vertical transport of oxygen on OH emission.

Thomas (1969) has a very interesting review of important airglow processes. To a large extent his paper suggests improvements to be made in the airglow observation program.

13.8 References

Armstrong, E. B., 1969. "Doppler shifts in the wavelengths of the OI-6300 lines in the night airglow." Planetary and Space Science 17, 957-974.

Abbott, W. N., 1964. "Possible correlations between sporadic-E and low-altitude airglow emission of 5577A in the median latitudes." Journal of Atmospheric and Terrestrial Physics, 26, 776-780.

Babcock, H. D., 1923. "A study of the green auroral line by the interference method." Astrophysical Journal 57, 209-221.

Baker, D. L. and R. O. Waddoups, 1967. "Rocket measurement of midlatitude night airglow emissions." Journal of Geophysical Research, 72, 4881.

Baker, D. L. and R. O. Waddoups, 1968. "Correction of paper by D. Baker and R. Waddoups, Rocket measurement of midlatitude night airglow emission." Journal of Geophysical Research, 73, 2546.

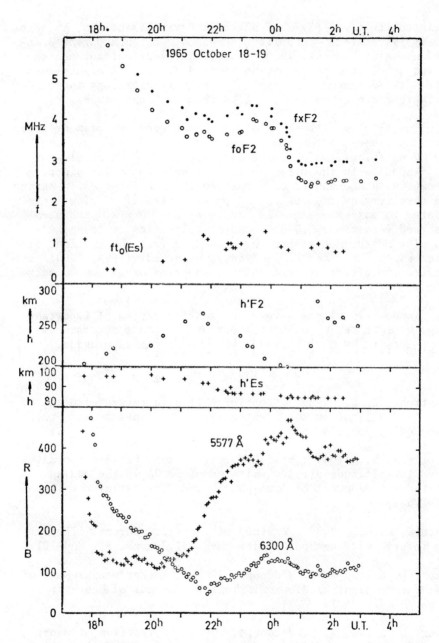

Fig. 13.8 Zenith intensities (in R) of the OI-5577 and the
OI-6300 nightglow emissions at Tsumeb as compared to ionospheric
virtual heights (h) in km, ordinary sporadic E top frequency,
ft_o (Es), and the critical F2 layer frequencies, for October
18 to 19, 1965 (from Dachs, 1969).

Barbier, D., 1959. "Systematic variations of the intensity of the principal radiations of the light of the night sky at the Haute-Provence observatory." Annals of Geophysics, 15, 412-414.

Barbier, D., 1961. "Les Variations d'intensite de la raie 6300 A de la luminescence nocturne." Annals of Geophysics, 17, 3.

Bates, D. R., 1960. "The airglow." Physics of the Upper Atmosphere, J. A. Ratcliffe, Editor, pp. 219, Academic Press, New York and London.

Bates, D. R. and M. Nicolet, 1950. "The photochemistry of the atmospheric water vapor." Journal of Geophysical Research, 55, 301-327.

Brenton, J. G. and S. M. Silverman, 1970. "A study of the diurnal variations of the 5577A (OI) airglow emissions at selected IGY stations." Planetary and Space Science, 18, 641-653.

Broadfoot, A. L., 1967. "Resonance scattering by N_2^+." Planetary and Space Science, 15, 1801.

Broadfoot, A. L., 1968. "Tropospheric Scattering of twilight N_2^+ emission." Planetary and Space Science, 16, 693.

Broadfoot, A. L. and D. M. Hunten, 1966. "N_2^+ emission in the twilight." Planetary and Space Science, 14, 1303-1319.

Chamberlain, J. W., 1961. Physics of the Aurora and Airglow, Academic Press, New York and London.

Chamberlain, J. W. and N. J. Oliver, 1953. "OH in the airglow at high latitudes." The Physical Review, 90, 1118.

Chapman, S., 1931. "Some Phenomena of the upper atmosphere." Proceedings of the Royal Society, (London), A 132, 353-374.

Chapman, S., 1937. "Production of auroral and nigh sky light." Philosophical Magazine, 23, 657-665.

Chapman, S., 1939. "Notes on atmospheric sodium." Astrophysical Journal 90, 309-316.

Christophe-Glaume, J., 1965. "Etude de la raie 5577A de l' oxygene dans la luminescence atmospherique nocturne." Annals of Geophysics, 21, 1-56.

Cole, K. D., 1965. "Stable auroral red arcs, sinks for energy of D_{st} main phase." Journal of Geophysical Research 70, 1689.

Cole, K. D., 1965. "The predawn enhancement of 6300A airglow." Annals of Geophysics 21, 156-158.

Dachs, J., 1967. Keinheubacker berickte, 12, 331.

Dachs, J., 1969. "On the nightglow intensity variations connected with vertical movements in the upper atmosphere." In Atmospheric Emission, B. M. McCormac and A. O. Omholt, Editors, pp. 515-517, Van Nostrand Reinhold Company, New York.

Deehr, C. S., 1968. "The twilight enhancement of the auroral and nebular lines of neutral atomic oxygen." Ph. D. Thesis, University of Alaska College, Alaska.

Fastie, W. S., 1968. "Far ultraviolet day airglow studies." Planetary and Space Science 16, 929-935.

Fedorva, N. I., 1967. "Hydroxyl emission at high latitudes." In Aurorae and Airglow No. 13" (Izd. Nauka, Moxcos), pp. 22-36.

Hesstvedt, E., 1969. "Airglow and vertical eddy transport." In Atmospheric Emissions, B. M. McCormac and A. Omholt, Editors, pp. 501-505, Van Nostrand Reinhold Company, New York.

Hesstvedt, E., 1970. "A theoretical study of the diurnal variation of hydroxyl emission." Journal of Geophysical Research 75, 2337.

Hunten, D. M., 1964. "Metalic emissions from the upper atmosphere." Science, 145, 26.

Hunten, D. M. 1967. "Spectroscopic studies of the twilight airglow." Space Science Reviews, 6, 493-473.

Hunten, D. M. and G. G. Shepherd, 1954. "A study of sodiu-in twilight II. Observations on the distribution." Journal of Atmospheric and Terrestrial Physics, 5, 57-62.

Kvifte, G. J., 1967. "Hydroxyl rotational temperatures and intensities in the nightglow." Planetary and Space Science, 15, 1515.

Kvifte, G. J., 1969. "Twilight observations", In Atmospheric Emissions, B. M. McCormac and A. Omholt, Editors, pp. 399-410, Van Nostrand Reinhold Company, New York.

Krasovskii, V. I., N. N. Shefov and V. I. Yarin, 1962. "Atlas of the airglow spectrum 3000 - 12,400 A." Planetary and Space Science, 9, 883-915.

McPherson and Vallance Jones, 1960. "A study of the latitude dependence of OH rotational temperatures for Canadian stations." Journal of Atmospheric Physics, 17, 302.

Maeda, L. R., 1962. Proceedings of the International Conferences on the Ionosphere, Physical Society, London, 187.

Megil, L. R. and N. P. Carleton, 1964. "Excitation by local electric fields in the aurora and airglow." Journal of Geophysical Research, 69, 101-122.

Meinel, A. B., 1948. "The near-infrared spectrum of the night sky and aurora." Publication of the Astronomical Society Pacific, 60, 373-378.

Meinel, A. B., 1950. "OH emission bands in the spectrum of the night sky, I." Astrophysical Journal, 111, 555-564.

Meinel, A. B., 1950. "OH emission bands in the spectrum of the night sky, II." Astrophysical Journal, 112, 120-130.

Meinel, A. B., 1950. "O_2 emission bands in the infrared spectrum of the night sky." Astrophysical Journal, 112, 464-468.

Packer, D. M., 1961. "Altitudes of the night airglow radiations." Annals of Geophysics, 17, 67-75.

Prokudina, V-S., 1959. "Observations of the line 6562A in the night airglow spectrum." In Spectral, Electrophotometrical and Radar Researches of Aurorae and Airglow, No. 1, pp. 43-44, Publishing House, Academy of Science, Moscow.

Prokudina, V. S., 1959, "Determination of the rotation temperature of hydroxyl in the upper atmosphere." Izv. Akad, Nauk, SSR, Ser. Geofiz., 125, 629.

Reay, N. K. and J. Ring, 1969. "Radial velocity and intensity measurements of the night sky H emission line." Planetary and Space Science, 17, 561-573.

Roach, F. E., 1968. "Photometric observations of the airglow during IQSY." In Geophysical Measurements: Techniques, Observational Schedules and Treatment of Data Annals of the IQSY, Vol 1, The M.I.T. Press, Massachusetts Institute of Technology, Cambridge, Massachusetts, and London, England.

Roach, F. E., 1969. "The nightglow and the F region ionosphere." In Atmospheric Emissions, B. M. McCormac and A. Omholt, Editors, pp. 439-447, Van Nostrand Reinhold Company, New York.

Roach, F. E. and L. L. Smith, 1967. "The worldwide morphology of the atomic oxygen nightglows". In Aurora and Airglow, B. M. McCormac, Editor, pp. 29-39, Reinhold Publishing Corporation, New York.

Roach, F. E., L. L. Smith and J. R. McKennan, 1969. "Nightglow observation during the IQSY." In Solar-Terrestrial Physics: Solar Aspects (Proceedings of Joing IQSY/COSPAR Symposium, London, 1967, Part I); Annals of IQSY, Vol. 4", the M.I.T. Press, Massachusetts Institute of Technology, Cambridge, Massachusetts and London, England.

Sidorov, V. N., 1968. "Some results of IQSY observations of hydroxyl emission." Aurorae and Airglow, No. 18, (Izd. Nauka, Moscow)

Silverman, S. M., 1969. "Night airglow phenomenology." In Atmospheric Emissions, B. M. McCormac and A. Omholt, Editors pp. 383-397, Van Nostrand Reinhold Company, New York.

Silverman, S. M., G. J. Hernandez, A. Carrigan and T. Markham, 1965. Handbook of Geophysics and space environments, S. L. Valley, Editor, McGraw-Hill, New York.

Slipher, V. M., 1929. "Emissions in the spectrum of the light of the night sky." Publication of the Astronomical Society Pacific, 41, 262-263.

Slipher, V. M., 1933. "Spectrographic Studies of the Planets." Monthly Note of the Royal Astronomical Society, 93, 657-668.

Stair, A. T. and H. P. Gauvin, 1967. "Research on optical infrared characteristics of aurora and airglow (artificial and natural)." In Aurora and Airglow, B. M. McCormac, Editor, Reinhold Publishing Corporation, New York.

Steiger, W. R., W. E. Brown and F. E. Roach, 1966. "The alignment of 6300A airglow isophotes in the tropics." Journal of Geophysical Research, 71, 2846.

Swider, W., Jr., 1969. "Upper atmospheric emission processes". In Atmospheric emissions, B. M. McCormac and A. Omholt, Editors, pp. 367-382, Van Nostrand Reinhold Company, New York.

Thomas, L., 1969. "Ionospheric phenomena related to airglow." In Atmospheric Emission, B. M. McCormac and A. Omholt, Editor pp. 423-434, Van Nostrand Peinhold Company, New York.

Tohmatsu, T. and T. Nagata, 1963. "Dynamical studies of the oxygen green line in the airglow." Planetary and Space Science, 10, 103.

Young, R. A. and G. Black, 1966. "Excitation of the auroral green line in the earth's nightglow." Planetary and Space Science, 14, 113-116.

CHAPTER 14

WHITE SANDS MISSILE RANGE METEOR TRAIL RADAR DESIGN

Alton A. Duff

Atmospheric Sciences Laboratory,
US Army Electronics Command,
White Sands Missile Range, New Mexico

14.1   Introduction

The Meteor Trail Radar (MTR) facility of the Atmospheric
Sciences Laboratory (ASL) at White Sands Missile Range (WSMR)
is a pulsed doppler system patterned after a unit built for the
Air Force Cambridge Research Laboratories (Ramsey and Myers,
1968), which in turn was similar to Greenhow's (1958) original
equipment developed in England during the 1950's. However, the
WSMR MTR incorporated several unique and significant changes
from these earlier designs to facilitate current applications.
It is designed to provide real-time range support and special
research data on component winds and air density in the 85-105
km altitude range by utilizing as the meteorological sensing
element the ionized trail left by meteors in that region of the
atmosphere.

The WSMR MTR operates at 32.8 MHz. It utilizes a monopulse antenna system for angular data acquisition. The MTR
facility is operated under digital controller (DC) synchroni-
zation and is programed so that regions of the atmosphere in
two orthogonal directions are searched alternately and nearly
simultaneously for meteor trails. The MTR has provision for
comparative calibration by use of an ancillary x-band tracking
radar and the simultaneous track of suitable targets of opportunity. Data are recorded on a pulse basis on magnetic tape
for research use, but the DC also averages the data for each
trail and outputs one line of real time information that represents the average data contained in the trail variations.

14.2   The MTR System

The MTR transmitts 32.8 MHz electromagnetic radiation at
50 kw peak power, 40 μs pulse width and at a rate of 500 pulses
per second. Energy is radiated in two component directions;

i.e., north and east, alternately (Fig. 14.1). For instance, pulse n is transmitted toward the north and pulse n+1 is transmitted toward the east. When an echo is received, the alternation of direction ceases, and all pulses are transmitted in the direction of the returned echo for a period of one-half second. The alternation of direction is then resumed.

Electromagnetic energy is reflected from the ionized trails which are produced by frictional interaction between high velocity extra terrestrial meteor particles and the low density atmosphere in the 85-105 km region. The reflected energy is received by a monopulse antenna array consisting of two groups of two antennas, each arranged above a common counterpoise (ground plane). Each group of two antennas receives the returned energy from its associated directional transmitting antenna (Fig. 14.2).

In each receiving group, one of the antennas is located one quarter wavelength and the other one-half wavelength above the counterpoise. The 43x43 meter counterpoise acts as a mirror to the electromagnetic energy returned from the ionized meteor trail. Each of the antennas receives and vectorially sums both the direct return and the returned energy reflected from the counterpoise. The output from each receiving antenna goes to a separate logarithmic receiver amplifier. By comparing these logarithmic receiver signal outputs in an additive circuit one obtains the ratio of the magnitude of the two signals; this ratio will be unique for any given elevation angle of arrival (Fig. 14.3).

The returned signal is further examined by the MTR electronics to determine the slant range, the rate of echo intensity decay, and the doppler shift of the frequency.

14.3 Data Handling System

The data and control words constitute some eight 16 bit digital words of raw echo data for each transmitted pulse. Two hundred and fifty transmitter pulses of data are taken for each echo (one-half second of data), and these data are stored between transmitter pulses in the digital controller's core memory in sequence.

After the one-half second data collection sequence has ended, the data are taken sequentially from core memory, properly formated to be Univac 1108 computer compatible, and recorded on magnetic tape in a Kennedy Model 1500 IR instrumentation recorder.

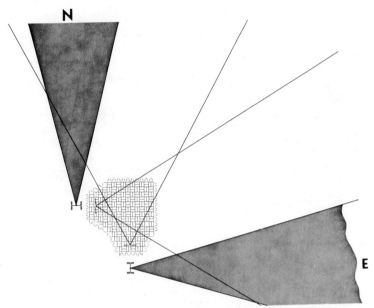

Fig. 14.1 Physical layout of the MTR antenna system. The hatched areas indicate approximate transmission patterns and the clear angles indicate approximate receiver patterns.

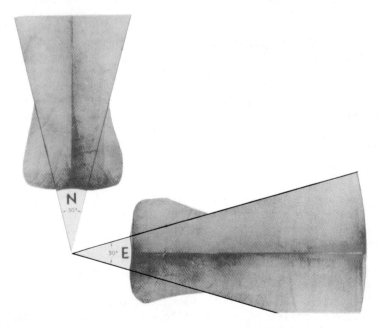

Fig. 14.2 Approximate horizontal intercept section.

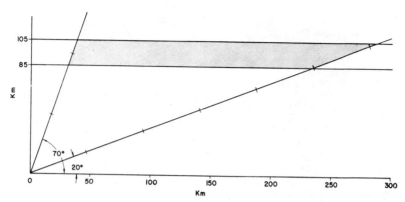

Fig. 14.3 Vertical intercept section.

The geometry of the atmospheric volume in which meteor trail radar wind measurements are made is illustrated in Fig. 14.3. It is clear that the mean positions of the zonal and meridional wind components are obtained at significantly different space locations.

The digital controller has been programed so that in case a second echo is received during the period required to output the data on magnetic tape, the output routing will be interrupted and another one-half second of data will be taken and stored in an alternate location of memory. The outputting of data on the magnetic tape and output writer will then be re-sumed and will continue until all stored data has been pro-cessed. If an echo is received while both memory locations have data in them, the new echo will be ignored. To take data and output it on magnetic tape and the IBM output writer takes approximately 30 seconds, so the MTR can handle about 120 echoes per hour based on internal equipment timing require-ments.

14.4 Calibration Equipment

One of the main problems associated with MTR observations is to maintain accurate and frequent calibration of the ele-vation determination antennas and circuitry. The WSMR MTR in-corporates a modified Nike-Hercules X-band tracking radar into the site complex, and comparative measurements of targets of opportunity are made by both the X-band and MTR radar system. When a target appears in the slant range area between a preset minimum range (to gate out ground clutter) and 85 km, the digital controller will recognize this as a calibration target, will slew the X-band track radar to the same azimuth quadrant,

elevation angle, and slant range, and will sound an alarm. As
soon as site personnel manually locate and track lock the X-band
radar onto the target, the MTR will start printing out comparative elevation angles at all azimuth angles of the track. Thus,
by the cumulative use of many targets of opportunity, the calibration table for the elevation angles can be improved and
corrections applied to the elevation computation program of
the digital controller as required.

The normal complement of standard test equipment for
circuit calibration is incorporated into the system. The electronics of this MTR are housed in a 2.5 x 12.2 meter trailer.
Two and one-half meters of the trailer length is partitioned
off, and this room contains the heating, air conditioning and
electrostatic precipitation air cleaning equipment. The remaining 9.8 meters of the trailer houses the electronics in
nine racks, equipment and file storage, technicians work
benches, and computation desk.

All trailers (MTR, site supply and maintenance, Nike-
Hercules track and Nike-Hercules computer-plot) are intercon-
nected with a 4.9 x 10 meter waning which will also serve as
site office space.

The WSMR MTR is designed and built as a research tool.
There can be a maximum of circuitry change with a minimum of
disruption to the research observational program. Future
modifications might include such things as pulse coded modulation for improved range accuracy, azimuth angle of arrival
determination, if needed, by the elevation sensing circuitry,
offset doppler for more accurate wind speed measurements, etc.

During the construction and checkout of the WSMR MTR it
became evident that the MTR system could be used as the nucleus
of an advanced type upper atmosphere sounding station. The MTR
would then not become an end in itself but would become a part
and extension of a complete meteorological station.

The background establishing the US Army's need for such
a system, the description of the equipment, the major modifications to existing instrumentation, and design concepts of
new equipment or techniques for this advanced upper atmosphere
meteorological sounding facility are given below.

Equipment and modifications to be incorporated in the advanced facilities include: omnidirectional meteor trail
tracking radar, a modified Nike-Hercules precision tracking
radar, and a control and data handling center to integrate the
meteorological data collected during operations. The MTR will

obtain wind and density data in the 85-105 km altitude range,
an approximate 40 km increase above normal rocketsonde observation altitudes. The Nike-Hercules precision radar is to be
extensively modified to become a Nike-Met radar and will be
used to establish calibration tables for the MTR antenna field
by simultaneously tracking targets of opportunity. It will
also track balloon and rocket payloads used in established
sounding programs. The time-shared digital controller
(Honeywell 516) will provide real-time output from the MTR,
rocketsonde, rawinsonde, and pibal operations along with ballistic computations and launcher setting outputs associated
with rocketsonde firings.

14.5 Advanced Meteor Trail Radar

The MTR to be fielded is a second generation, pulsed
doppler model designed by ASL and incorporates certain improvements over the system previously used.

The transmitting and receiving antennas are combined and
consist of a radome-enclosed, all weather, mechanically stable,
stacked, crossed dipole antenna with a vertical whip, centered
on and located above a counterpoise. The antenna, with
properly phased and amplitude controlled input through a
duplexer assembly, transmits an omnidirectional, in the azimuth
plane, inverted, approximately cosecant-squared pattern. In
the receiving mode, again with proper phasing, the antenna and
associated receivers provide multiple output voltages that,
with controller processing, result in the determination of
elevation and azimuth signal arrival angles. The system feeds
one receiver from each of the four dipoles in the stacked cross,
and the vertical whip antenna has its own receiver. It is
evident that the five receivers must be closely matched in
phase, gain, and gain tracking through a wide range of received
signal strengths. The improved monopulse antenna system thus
provides for omnidirectional operation of the MTR between elevation angles of 20 to 70 degrees in lieu of scanning only a
small volume of the surrounding atmosphere.

An additional improvement provided by this system is the
shaping of the transmitter RF output pulse by RF phase modulation techniques (Thomas and Ward, 1969). The sidelobe
structure of the transmitted frequency spectrum is greatly
reduced, and operation in an already crowded band around 32.8
MHz is more easily accomplished with less radio frequency
interference. In this technique, the RF transmitter pulse is
composed of two equal-amplitude and equal-shaped pulses occurring exactly in time phase. The RF phase of one of the pulses
is smoothly shifted throughout its period. Thus, at the start

and at the end of the pulse, its RF phase differs by 180
electrical degrees from the reference, while in the center
portion of the pulse the two signals are in RF phase. The two
signals are then combined in a hybrid junction to form the
final transmitted pulse. A variation of this technique, where
the RF phase of both pulses is changed, can produce phase-coded
information within a long RF pulse (Benjamin, 1966). This
coded information can be computer processed to increase the
accuracy of range determination of the meteor trail.

14.6 Advanced Digital Controller

The digital controller serves as the master synchronizer
for all operations on the station, including the timing syn-
chronization of all portions of the MTR, furnishing the master
synchronizer pulse for the Nike-Met radar system, and controlling all data collection, processing, and recording functions.
The controller is programed to analyze data signals and to
make decisions according to a predetermined order of priority
concerning the type of data collection mode in which the station
should be operated. For example, a target of opportunity will
prescribe a certain tracking mode. Targets of opportunity are
defined as those located between ground clutter range and 85 km
or those that persist longer than a predetermined length of
time. The Nike-Met radar will then slew to this target's
azimuth, elevation angle, and range and sound an alarm. When
the alarm sounds, site personnel track lock the Nike-Met radar
onto the target. The controller compares the very accurate
track data from the Nike-Met with that from the MTR. If necessary, the controller automatically updates the antenna calibration tables of the MTR for more precise MTR observations.

14.7 Nike-Met Radar

Technical considerations and limited availability of
adequate tracking radar systems led to the selection of the
Nike-Hercules X-band tracking radar as the precision radar for
the facility. Modifications to optimize the radar for its in-
tended use as the Nike-Met X-band precision tracking radar include:

1. Replacement of the present antenna system with a 12-
foot, fiberglass-covered foam, Cassegrain antenna system by
raising the elevation axis trunion bearings on the yoke.

2. Incorporation of a three-channel X-band parametric
amplifier in the forward end of the Nike-Met radar system.

3. Mounting of a 1680 MHz monopulse antenna feed system
around the X-band antenna feed for compatibility with the

meteorological telemetry system (addition of this feature will permit tracking in the X-band and the 1680 MHz band simultaneously with the reception of telemetry on the 1680 MHz band antenna feed).

    4. Inclusion of a parametric RF amplifier in the 1680 MHz band antenna feed system for maximum signal-to-noise ratio of the received telemetry signal.

    5. Use of gray binary code plates on the azimuth and elevation axes of rotation of the antenna pedestals to encode angular data at the point of generation. Associated circuitry for data transfer to the MTR controller; encoding of range data in a one nanosecond resolution counter system for the controller.

    6. Sampling of the data output from the Nike-Met radars at a 10 per second rate and recording of this information in real-time on magnetic tape.

    7. Sample-and-hold and analog-to-digital conversion devices on normal video, automatic gain control, and three axes of tracking error voltages. (The information thus obtained will be combined in the digital controller with the code plate digital track information to provide more accurate total track data by the electrical elimination of mechanical tracking errors.)

    8. To simplify operations, the modified system provides a display of tracking angles and other operational data of both radar pedestals on each "A" scope screen of the Nike-Met radar along with the tracking video as shown in Fig. 14.4. The last confirmed location of the target is displayed so that, if the target track is lost, the operator will have the best "look" angles and range data available for reacquisition. For maximum help in acquisition, the controller can be programed prior to a rocket firing to display extrapolated look angles in real-time after the time of firing and, if desired, the Nike-Met radar can be kept slewed to this extrapolated search point.

    9. The Nike-Met radar is equipped with a subharmonic, pulsed transmitter and an antenna system piggy-backed and aligned with the 12-foot primary parabolic antenna with a double-the-frequency automatic frequency control (AFC) for the X-band receivers to permit the use of a two-frequency radar scheme. In this mode of operation, a pilot balloon being tracked by the radar is equipped with 4.5 GHz harmonic generating diodes in a one-half wavelength dipole wire antenna. The X-band radar transmitter is turned off, and the subharmonic transmitter emits short pulses of RF energy at 4.5 GHz. When

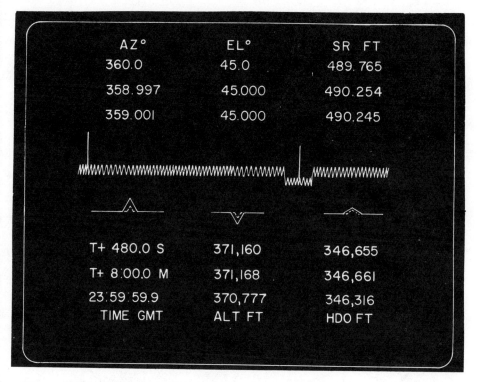

Fig. 14.4 Nike met "A" scope video display.

the signal is received by the diode assembly on the balloon, it is retransmitted at twice the frequency by harmonic generation. The 9.0 GHz signal is received and tracked by the X-band Nike-Met antenna and ranging system. Since the AFC locks the receivers onto a frequency of twice the transmitter frequency, the balloon is detected at 9.0 GHz and tracked in a background free of all radar ground clutter. Tracking can occur at ranges less than those dictated by the usual recovery times of the T-R tubes (which are not used in this configuration).

The Nike-Met radar pedestals are enclosed in heated, insulated, metal space frame radomes so that all-weather maintenance and operation of the equipment can be accomplished without loss of tracking accuracy as shown in Fig. 14.5.

14.8 Facility Layout

The Nike-Met radar track and plot electronics racks are removed from their trailers and placed with the MTR electronics racks on a floating floor in an insulated, all dielectric

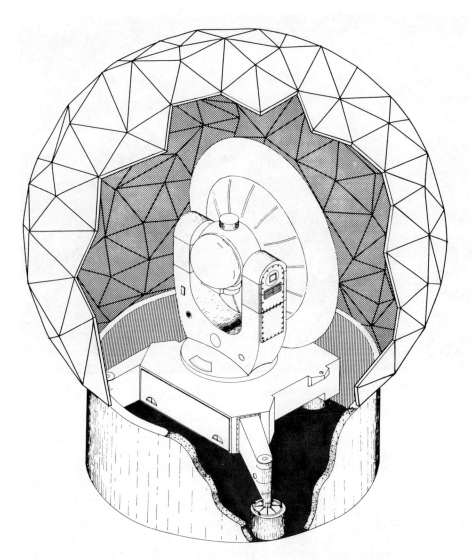

Fig. 14.5   Nike-Met tracking radar and insulated radome.

radome 18 meters in diameter as shown in Fig. 14.6. The ceiling
of the equipment room in the radome is placed at the equatorial diameter of the radome.   It is made of metal-faced, 10 cm
polyurethane foam insulation structure board and forms the
solid metal center part of the MTR counterpoise.   The counterpoise, extending from the radome some 20 meters along the
radius, is composed of 15 centimeter square electrically conducting mesh properly supported and leveled.   The Nike-Met
radar track pedestals in their insulated radomes are located

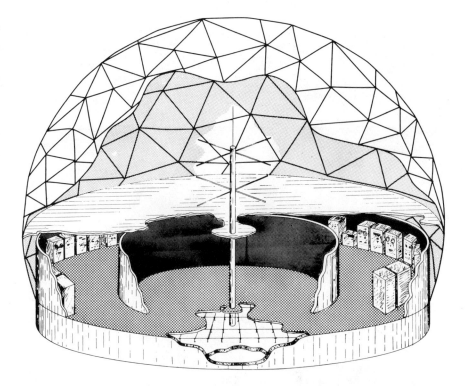

Fig. 14.6.   Meteor trail and Nike-Met radar equipment rack layout.

beyond the edge of the counterpoise on a diameter through the center radome to reduce line-of-sight interference.

In areas where the sounding facility provides its own site air surveillance (such as the Alaska station) a TPS-1D surveillance radar is incorporated into the station; however, where air surveillance is provided by the range host (Fort Sherman, Canal Zone), only the MTR and the Nike-Met X-band precision tracking radars are used. A typical facility layout for a remote station (as in Alaska) is shown in Fig. 14.7.

14.9   Conclusion

When installed in the summer of 1971 and completed in 1972 (funds permitting), these modified sounding facilities will provide upper atmosphere meteorological sounding support at remote sites comparable to the best missile range measurements, with a significant reduction in expenditure of material and personnel.

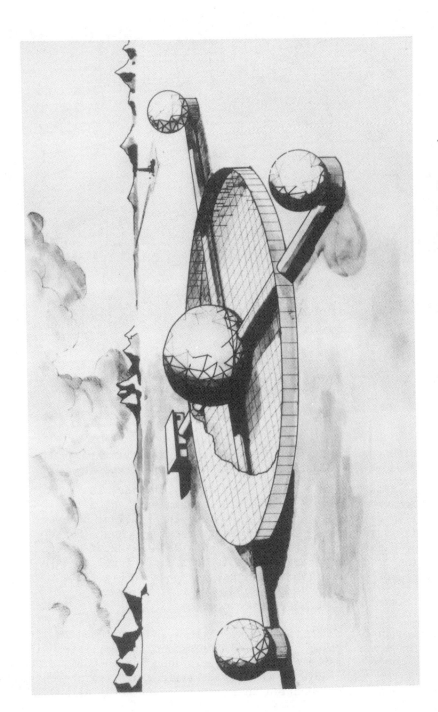

Fig. 14.7   Typical site layout of the meteor trail and Nike-Met radars.

14.10   References

Benjamin, R., 1966. Modulation, Resolution and Signal Processing in Radar, Sonar and Related Systems. Pergamon Press, Oxford, 184 pp.

Greenhow, J. S., 1958. "Meteor Trail Measurements by Radio Echo Detection Means," University of Manchester, Jodrell Bank Experimental Station, Technical Note No. 2, 25 pp.

Ramsey, W. H., and R. F, Myers, 1968. "The AFCRL Pulse Doppler Radar for the Determination of Winds and Density from Master Trails." In Proceedings of the Workshop on Methods of Obtaining Winds and Densities from Radar Meteor Trail Return, pp. 39-65, edited by A. A. Barnes and Joseph J . Pazniokas, Air Force Cambridge Research Laboratories, AFCRL-69-0228, L. G. Hanscom Field, Bedford, Mass.

Thomas, H. J., and Ward, H. R., 1969. Study and Experimentation of RF Pulse Generating and amplifying Techniques for Improving Radar Emission Characteristics, Ratheon Company, Wayland Laboratories, Wayland, Mass., Final Report No. RD-69-33.

CHAPTER 15

METEOR TRAIL RADAR DATA PROCESSING

Bruce T. Miers

Atmospheric Sciences Laboratory,
US Army Electronics Command,
White Sands Missile Range, New Mexico

15.1   Introduction

The first measurements of wind velocity in the region of 90 km by means of radar reflections from meteor trails were made at Stanford University by Manning et al. (1949). Subsequently other investigators have constructed meteor trail radars of various types, but no formal data exchange of individual measurements has resulted. While results of several studies of atmospheric structure from meteor trail data have been published in various journals, the measurements obtained through use of these systems generally remains unavailable for study by individual scientists conducting upper atmospheric research.

A rather excellent example of an optimum mode in which cooperative action can be implemented by individual researchers for synoptic studies, and thus the individual measurements value notably enhanced, is illustrated by the Meteorological Rocket Network (MRN) (Webb, 1966, 1969; Ballard, 1967) which has facilitated synoptic exploration of the stratospheric circulation during the past decade. These data exchanges contain only those data suitable for synoptic exploration of the upper atmosphere, and thus leaves the primary investigator with the bulk of the acquired data for conduct of his individual research programs. It is proposed here that the formats presented below be used for immediate dissemination of a limited amount of MTR data to the scientific community for synoptic exploration of the dynasphere.

15.2   Immediate Data Exchange Format

Each MTR station should make available to the scientific community a specified amount of the acquired data which will optimumly indicate the synoptic situation in the region of the atmosphere which it observes. These data are proposed to be

two hour average of the observed mean wind measurements in the
90-100 km altitude region, to be designated the Thermospheric
Circulation Index (TCI), analogous to the Stratospheric Circulation Index (SCI) of the MRN synoptic exploration effort.
The TCI should be presented in zonal and meridional components,
using the usual meteorological directional nomenclature with
speeds in mps.

The format for data transmission should be

$$S_i S_i S_i S_i S_i \quad MMDDY \quad TTTTZ \quad GWWWW \quad GWWWW,$$

where the first group is World Meteorological Organization
station identifier, the second group are the month, day and
year, the third group is the time of the middle of the observational period given in Greenwhich Mean Time, the fourth group
is the north-south wind component in mps with the sign first
given (a wind from the north to south is designated by a minus
sign), the fifth group is the east-west wind component in mps
with the sign given first (a wind from the east to west is
designated by a minus sign).

15.3 Data Publication Format

After suitable data screening, variable according to the
characteristics of the particular radar system, the following
data format is recommended for unified recording and comparison
of data obtained from individual meteor trail observations.

Form   Data
Columns

1-5   Reporting point: A five digit WMO Code number for
          the particular location
6   Blank
7-8   Year: The last two digits of the year are entered.
9-10   Month: January is entered as 01; December as 12.
11-12   Day: Days are entered as 01 to 31 where appropriate.
13-18   Time: Time of the meteor trail is shown in hours,
          minutes and tenths of minutes Greenwich Mean Time,
          based on a 24-hour clock. A decimal is recorded in
          Column 17.
19   Blank
20-25   Trail Number: Meteor trails will be numbered con-
          secutively for each day beginning with the first
          trail after 0000.0 GMT
26-28   Range: The range in whole kilometers from the radar
          to the meteor trail.
29-33   Decay: The decay rate of the echo in dB/sec.

Form Columns	Data
34-36	Azimuth Angle: The angle is given in whole degrees with 000 being true north, 090 east, 180 south and 270 west.
37-40	Elevation Angle: The angle is given in degrees and tenths of degrees with a decimal in Column 39.
41	Blank
42-45	Doppler Information: Speed of the doppler shift in whole mps. Column 42 will have a plus (+) if toward the station and a (-) if away from the station.
46-50	Density Height: The density height is computed from the decay rate and the standard atmosphere and recorded in kilometers and tenths of kilometers. A decimal is recorded in Column 49.
51	Blank
52-57	Wind Direction and Speed: The wind components will be given in mps in Columns 54-57, where, in Columns 52 and 53, a 6- indicates a wind from the north and 7+ indicates a wind from the south, an 8- indicates and east wind and a 9+ indicates a west wind (i.e.,: 6.9125 in the columns means a wind from the north at 125 m sec^{-1}).
58-60	Standard Deviation of the Wind: The standard deviation of the wind is given in mps.
61-63	Height: The height of the given wind in whole kilometers. The wind values will be averages over a one kilometer thick layer. The height is computed from the slant range and elevation angle.
64	Blank
65-68	Special Case Wind Direction: If the radar orientation is such that the usual component winds are not possible, the direction will be given in these columns (i.e., 878 indicates a wind from the ESE, with columns 52 and 53 above left blank).
69-80	Blank

When certain data provided for on the form are not recorded or are unavailable, the appropriate sections are left blank, and when only a portion of a particular section is filled with a number or numbers, the remainder of this section is filled with zeros. Missing data are left blank.

15.4   References

Ballard, H. N., 1967. "A Guide to Stratospheric Temperature and Wind Measurements." COSPAR Technique Manuel Series, COSPAR Secretariat, 55 Boulevard Malesherbes, Paris 8e, France, 117 pp.

Manning, L. A., 1949. "Radio Doppler Investigation of Meteoric Heights and Velocities," Journal of Applied Physics, Vol. 20, pp. 475-479.

Webb, W. L., 1966. Structure of the Stratosphere and Mesosphere, Academic Press, Inc., New York, 382 pp.

Webb, W. L., 1969 (editor). AIAA Progress in Astronautics and Aeronautics: Stratospheric Circulation, Vol. 22, Academic Press, Inc., New York, 600 pp.

Index to
Contributors to Volume 27

Bruce, Rufus E.
UNIVERSITY OF TEXAS AT EL PASO
331

Duff, Alton A.
U. S. ARMY ELECTRONICS COMMAND
355

Fogle, Benson
NATIONAL CENTER FOR ATMOSPHERIC
RESEARCH
95

Grossi, M. D.
SMITHSONIAN INSTITUTION
ASTROPHYSICAL LABORATORY
205

Haurwitz, B.
NATIONAL CENTER FOR ATMOSPHERIC
RESEARCH
109

Heikkila, Walter J.
THE UNIVERSITY OF TEXAS AT DALLAS
53

Hesstvedt, Eigil
UNIVERSITY OF OSLO
327

Hines, Colin O.
UNIVERSITY OF TORONTO
79

Midgley, James E.
UNIVERSITY OF TEXAS AT DALLAS
37

Miers, Bruce T.
U. S. ARMY ELECTRONICS COMMAND
369

Nowak, R.
STANFORD UNIVERSITY
249

Peterson, A. M.
STANFORD UNIVERSITY
249

Roper, R. G.
GEORGIA INSTITUTE OF TECHNOLOGY
181

Rosenthal, S. K.
SMITHSONIAN INSTITUTION
ASTROPHYSICAL LABORATORY
205

Sechrist, C. F., Jr.
UNIVERSITY OF ILLINOIS
261

Southworth, R. B.
SMITHSONIAN INSTITUTION
ASTROPHYSICAL LABORATORY
205

Spizzichino, Andre
CENTRE NATIONAL D'ETUDES DES
TELECOMMUNICATIONS, FRANCE
117

Webb, Willis L.
U. S. ARMY ELECTRONICS COMMAND
1